Microemulsion - a Chemical Nanoreactor

Edited by Juan C. Mejuto

Published in London, United Kingdom

IntechOpen

Supporting open minds since 2005

Microemulsion - a Chemical Nanoreactor
http://dx.doi.org/10.5772/intechopen.73338
Edited by Juan C. Mejuto

Contributors
Concha Tojo, M.Arturo López-Quintela, David Buceta, Tomás Viveros-García, Jose Luis Munguía-Guillén, José Antonio De Los Reyes-Heredia, Michel Picquart, Marco Antonio Vera-Ramírez, Juan C. Mejuto, Antonio Cid, Manuel Alonso-Ferer, A Acuña, Gonzalo Astray, Luis García-Rio, Jesus Simal-Gandara, Pedro Rodríguez-Dafonte, Rohini Kanwar, S.K. Mehta, Jyoti Rathi, Madhuri Patil, Mysara Mohyaldinn, Anas Hassan, Mohammed Abdalla Ayoub Mohammed

First published in London, United Kingdom, 2019 by IntechOpen
IntechOpen is the global imprint of INTECHOPEN LIMITED, registered in England and Wales, registration number: 11086078, The Shard, 25th floor, 32 London Bridge Street
London, SE19SG – United Kingdom
Printed in Croatia

British Library Cataloguing-in-Publication Data
A catalogue record for this book is available from the British Library

Additional hard and PDF copies can be obtained from orders@intechopen.com

Microemulsion - a Chemical Nanoreactor
Edited by Juan C. Mejuto
p. cm.
Print ISBN 978-1-78984-501-3
Online ISBN 978-1-78984-544-0
eBook (PDF) ISBN 978-1-78984-545-7

Meet the editor

Juan Carlos Mejuto is Full Professor in the Physical Chemistry Department of the University of Vigo at the Ourense Campus. He is the head of the Colloids group at Ourense Campus. His research interest comprises physical organic and physical inorganic chemistry, reactivity mechanisms in homogeneous and microheterogeneous media, stability of self-assembly aggregates, and supramolecular chemistry.

Contents

Preface

In Memoriam, Prof. Julio Casado

Microemulsions are self-assembled colloidal aggregates formed by a dispersed phase and a continuous phase stabilized by a surfactant. These aggregates may or may not require the presence of a fourth component known as a cosurfactant.

These systems are of great interest from a chemical point of view because they facilitate the simultaneous presence of a hydrophobic environment and a hydrophilic environment, with which they have an important solubilizing capacity, distributing the solutes between the water, the organic phase, or in surfactant film depending on their physicochemical characteristics. For this reason, they have abundant and interesting applications in different fields such as solubilization or extraction. Likewise, depending on their biocompatibility, they can be used as vehicles for active principles and drugs, improving their adsorption and therefore their efficacy from a pharmacological point of view.

Therefore, and throughout this book, we will see important applications of microemulsions as chemical nanoreactors. In the first place, they can behave as microheterogeneous catalysts, where one can observe both an important catalysis, an effect of increasing the concentration of reagents in one of the domains of the system, and a perfect inhibition, due to the physical separation of the reactants when solubilized in different phases of the system. Thus, in the literature there are a large number of examples where microemulsions have been used as reaction media in chemical, photochemical, and enzymatic processes. Even microemulsions have been used as an alternative system to phase transfer catalysts.

In addition, the diameter of the microdroplets that form the dispersed phase can be modulated, so that the nanoreactor can be scalable. In this sense, microemulsions can be used to control the growth of nanocrystals or the diameter of nanoparticles.

We must not forget the fact that they can be used to modulate mechanisms oriented to stereoselective synthesis.

Last, but not least, it is necessary to underline a final characteristic: their ability to simulate complex biological structures. This makes microemulsions a valuable tool for biomimetic chemistry. Thus, microemulsions have been used as models of biological membranes or as models of water trapped in restricted media.

Leaving aside the chemistry, at the same time that the adventure of editing this book began, on July 2, 2018, Professor Julio Casado died in Salamanca. For this reason, I want to use this prologue to pay tribute. Don Julio Casado-Linarejos was born in Palencia in 1939 and received his PhD in Chemistry from the University of Valladolid in 1965. After the defense of his doctoral thesis, he joined as a postdoctoral researcher with Professor Bak at the University of Copenhagen, establishing on his return the first microwave laboratory in Spain.

In 1973, at only 34 years of age, he was appointed Professor at the University of Santiago de Compostela (Galicia), where he formed a research group on chemical kinetics that has trained a large part of chemistry teachers in the university district of Galicia, constituting a school to which I have the honor of belonging as one of his disciples. Within this research group, the chemistry of nitroso compounds was studied in a very detailed way, both in aqueous and non-aqueous media. From these studies, and under the direction of Professor Leis and Professor Elena Peña, we addressed the study of the chemistry of these compounds in microheterogeneous media. In Santiago de Compostela he was also Vice-Rector of the University and Director of the Institute of Educational Sciences. Later, he moved to the University of Salamanca where he finished his research career, first as a professor and then as emeritus, and not without leaving an important imprint among the Galician physical chemists.

In his curriculum, it should be noted that he received the Medal of the Real Spanish Society of Chemistry and the Prize of Investigation Aldrich Chemistry. He was named Doctor Honoris Causa by the University of Vigo, where I had the honor of being his godfather. In addition, I occupy the direction of the Spanish National Agency of Evaluation and Prospective.

This old Castilian by birth, but Galician by adoption, was, above all, a citizen of the world, who not only taught us chemistry but also to be better people. He was a man who always related with sympathy to people regardless of their social status or political ideology, and who always knew how to see goodness in the human being; he left a deep impression on all those who had the good fortune to know him.

His deep culture turned him into a true Renaissance man who devoted his life to teaching science with rigor and, above all, with love.

Sit tibi terra levis

Juan C. Mejuto
Physical Chemistry Department,
University of Vigo,
Spain

Section 1

Chemical Kinetics and Microheterogeneous Catalysis

Chapter 1

Pseudophase Model in Microemulsions

Antonio Cid, Aangel Acuña, Manuel Alonso-Ferrer, Gonzalo Astray, Luis García-Río, Jesus Simal-Gándara and Juan C. Mejuto

Abstract

The kinetic behaviours in microemulsions can be easily modelled using an extension of the pseudophase model previously developed for micellar catalysis. This model considers that the microheterogeneous media can be considered as the sum of different conventional reaction media, where the reagents are distributed and in which the reaction can occur simultaneously. The reaction rate observed in the microheterogeneous system will be the sum of the velocities in each one of the pseudophases. This use can be considered as an extension of the pseudophase model, which has been developed for the quantitative analysis of nitrosation reactions in AOT/isooctane/water microemulsions and has been applied successfully in the literature in a large variety of chemical reactions.

Keywords: microemulsions, reverse micelles, kinetic model, pseudophase model

1. Introduction

The use of microemulsions, in particular, colloidal self-aggregates, and in general, as reaction media [1] makes the application of kinetic models necessary for the quantitative interpretation of the observed results. In this sense, a simple thermodynamic model was developed for its application in micelles [2] and it was called the pseudophase model [3]. This model was successfully applied through extensions to different microheterogeneous systems over the last 50 years [4–10].

2. The pseudophase model

This model considers that the micellar system can be considered as the sum of two conventional reaction media, the continuous pseudophase and the micellar pseudophase, where the reagents are distributed and in which the reaction can occur simultaneously (see **Figure 1**).

In this figure, A and B are the reagents, k is the kinetic constant and the subscripts w and m denotes the reaction *loci* (w corresponds to the continuous pseudophase and m is the micellar pseudophase). Finally, K_A and K_B are the distribution coefficients of both reactants between different pseudophases.

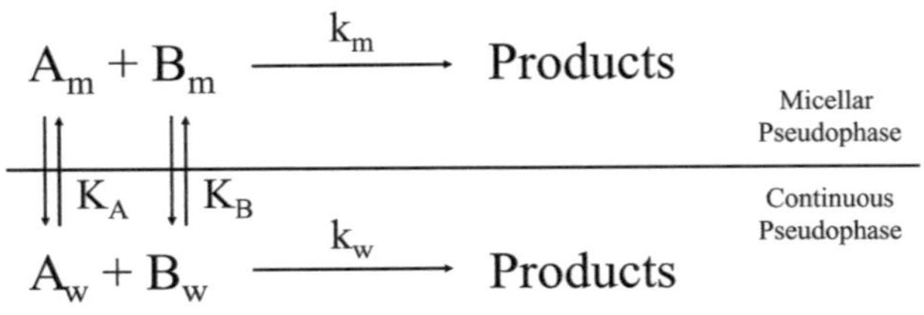

Figure 1.
Pseudophase model applied to micellar aggregates as reaction media.

The model considers that the reaction rate observed in the microheterogeneous system will be the sum of the velocities in each one of the pseudophases, and it can be expressed as shown in the following equations assuming a first-order reaction for each reactant:

$$v_T = v_w + v_m \quad (1)$$

$$v_T = k_{obs}[A_T][B_T] \quad (2)$$

$$v_w = k_w[A_w][B_w] \quad (3)$$

$$v_m = k_m[A_m][B_m] \quad (4)$$

$$k_2[A_T][B_T] = k_w[A_w][B_w] + k_m[A_m][B_m] \quad (5)$$

The mass balance on both pseudophases and the consideration of the distribution coefficients between them allow us to establish the existing relationship between the total concentrations of the reactants and the concentrations in each of the pseudophases considered.

$$[A]_T = [A]_w + [A]_m \quad (6)$$

$$[B]_T = [B]_w + [B]_m \quad (7)$$

$$K_A = \frac{[A]_m}{[A]_w[D_n]} \quad (8)$$

$$K_B = \frac{[B]_m}{[B]_w[D_n]} \quad (9)$$

In Eqs. (8) and (9), D_n denotes the concentration of surfactant micellized and $[D_n]$ = [D]-CMC (where *D* is the total surfactant concentration and *CMC* is the critical micellar concentration). Using Eqs. (8) and (9), the following equations can be written:

$$[A]_m = K_A[A]_w[D_n] \quad (10)$$

$$[B]_m = K_B[B]_w[D_n] \quad (11)$$

$$[A]_T = [A]_w + K_A[A]_w[D_n] \quad (12)$$

$$[B]_T = [B]_w + K_B[B]_w[D_n] \quad (13)$$

$$[A]_w = \frac{[A]_T}{1 + K_A[D_n]} \quad (14)$$

$$[\mathrm{B}]_{\mathrm{w}} = \frac{[B]_T}{1 + K_B[\mathrm{D}_n]} \tag{15}$$

$$[\mathrm{A}]_m = \frac{[A]_T K_A[\mathrm{D}_n]}{1 + K_A[\mathrm{D}_n]} \tag{16}$$

$$[\mathrm{B}]_m = \frac{[B]_T K_B[\mathrm{D}_n]}{1 + K_B[\mathrm{D}_n]} \tag{17}$$

Using Eqs. (14)–(17) on Eq. (5), the following expressions can be deduced (Eq. (18)–(21)):

$$k_2[A_T][B_T] = k_w \frac{[A]_T}{1 + K_A[\mathrm{D}_n]} \frac{[B]_T}{1 + K_B[\mathrm{D}_n]} + k_m \frac{[A]_T K_A[\mathrm{D}_n]}{1 + K_A[\mathrm{D}_n]} \frac{[B]_T K_B[\mathrm{D}_n]}{1 + K_B[\mathrm{D}_n]} \tag{18}$$

$$k_2[A_T][B_T] = k_w \frac{1}{1 + K_A[\mathrm{D}_n]} \frac{1}{1 + K_B[\mathrm{D}_n]} [A_T][B_T] + k_m \frac{K_A[\mathrm{D}_n]}{1 + K_A[\mathrm{D}_n]} \frac{K_B[\mathrm{D}_n]}{1 + K_B[\mathrm{D}_n]} [A_T][B_T] \tag{19}$$

$$k_2 = k_w \frac{1}{1 + K_A[\mathrm{D}_n]} \frac{1}{1 + K_B[\mathrm{D}_n]} + k_m \frac{K_A[\mathrm{D}_n]}{1 + K_A[\mathrm{D}_n]} \frac{K_B[\mathrm{D}_n]}{1 + K_B[\mathrm{D}_n]} \tag{20}$$

$$k_2 = k_w \frac{1}{(1 + K_A[\mathrm{D}_n])(1 + K_B[\mathrm{D}_n])} + k_m \frac{K_A K_B[\mathrm{D}_n]^2}{1 + K_B[\mathrm{D}_n]1 + K_A[\mathrm{D}_n]} \tag{21}$$

According to the pseudophase model, each pseudophase is evenly distributed in the total micellar dispersion volume. The value of rate constanst must be corrected taking into account the molar volume of each pseudophase to compare de intrinsic reactivity in the two different domains due to this reactants distribution between both pseudophases [10].

Equation (21) can be simplified according to the distributions of A and B, and the presence of chemical reaction in one or both pseudophases. This model predicts the catalysis or inhibition processes with success due to the compartmentalising effect of these colloidal aggregates [11–28]. However, this model must be expanded to take into account possible ion exchange processes between the continuous medium and the micelle, which give rise to more complicated expressions [29]. In some cases, it is necessary to resort to Poisson-Boltzmann distribution to evaluate the concentration of the different ions in the Stern and Gouy-Chapman layers to be able to model the ion exchange process between the continuous medium and the micellar electric double layer [30].

The pseudophase model applied to micelles has also been satisfactory for the analysis of the kinetic results in more complex micellar systems such as mixed micellar-cyclodextrin systems [31–37] or pseudo-micellar humic acids aggregates [38–42].

3. The pseudophase model in microemulsions

The pseudophase model was first extended by our research group in order to quantitatively analyse the kinetic behaviour of nitrosation reactions in microemulsions based on AOT [43, 44]. Afterwards, this extended model, with minor corrections, has been satisfactorily tested on microemulsions covering all possible cases [45–55] such as: (i) different chemical reactions (ionic or non-ionic),

(ii) reactants distributed throughout the different hydrophobic domains or
(iii) with different reaction loci.

Unlike in normal micelles, where we recognised two different domains (micelles and bulk water), in a microemulsion system three domains can be found: (i) the microdroplets of the dispersed phase, (ii) the continuous phase and (iii) the surfactant film (or surfactant + cosurfactant) that stabilises the system. Due to this, in this case, three pseudophases will be considered, taking into account the same proposed considerations for the micellar model.

We will assume that the reactants can be located in each of these three pseudophases, and their distribution will be governed by the distribution coefficients defined in an analogous way to that proposed in micelles. The chemical reaction can take place in each of the three pseudophases. In this way, the model can be explained according to **Figure 2**, where A and B are the reagents, k corresponds to the kinetic constant and the subscripts d, i and c denotes the loci (c corresponds to the continuous pseudophase, d with the dispersed pseudophase and i with the surfactant film -or interphase-). $K_{A,id}$ and $K_{B,id}$ are the distribution coefficients of both reactants between the interphase and the dispersed phase. $K_{A,ic}$ and $K_{B,ic}$ correspond to distribution coefficients of both reactants between the interphase and the continuous phase.

As in the case of micelles, the reaction rate observed in the microemulsions will be the sum of the velocities in each one of the pseudophases as it shown in the following equations (as in the case of micelles *-vide supra-*, assuming a first order reaction for each reactant):

$$\mathrm{v_T} = \mathrm{v_c} + \mathrm{v_i} + v_d \tag{22}$$

$$\mathrm{v_T} = k_{obs}[A_T][B_T] \tag{23}$$

$$\mathrm{v_d} = k_d[A_d][B_d] \tag{24}$$

$$\mathrm{v_i} = k_i[A_i][B_i] \tag{25}$$

$$\mathrm{v_c} = k_c[A_c][B_c] \tag{26}$$

$$k_2[A_T][B_T] = k_d[A_d][B_d] + k_i[A_i][B_i] + k_c[A_c][B_c] \tag{27}$$

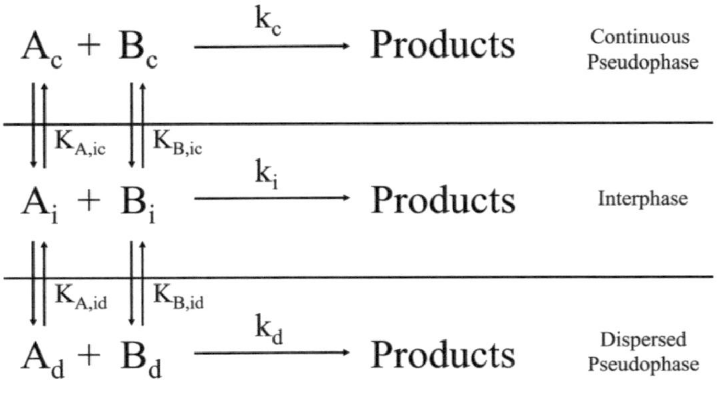

Figure 2.
Pseudophase model applied to microemulsions reaction media.

As quoted above, the mass balance on the three pseudophases and the consideration of the distribution coefficients between them allow us to establish the existing relationship between the total concentrations of the reactants and the concentrations in each of the pseudophases considered

$$[\mathrm{A}]_\mathrm{T} = [\mathrm{A}]_\mathrm{c} + [\mathrm{A}]_\mathrm{i} + [\mathrm{A}]_\mathrm{d} \tag{28}$$

$$[\mathrm{B}]_\mathrm{T} = [\mathrm{B}]_\mathrm{c} + [\mathrm{B}]_\mathrm{i} + [\mathrm{B}]_\mathrm{d} \tag{29}$$

$$K_{A,\,id} = \frac{[A]_i[Dis]}{[A]_d[D_n]} \tag{30}$$

$$K_{B,\,ic} = \frac{[B]_i[C]}{[B]_w[D_n]} \tag{31}$$

where [C] is the continuous phase concentration and $[D_n]$ corresponds to the concentration of surfactant in the microemulsion. In the case of micelles, [Dn] is obtained as $[D_n]$ = [D]-CMC; but in the case of microemulsions, CMC = 0. It means that the surfactant concentration in the microemulsion is equal to the total surfactant concentration. Similar expressions can be obtained for the partition coefficients between the dispersed pseudophase and the interphase. In this case, [Dis] corresponds to the dispersed phase concentration.

$$K_{A,\,id} = \frac{[A]_i[Dis]}{[A]_d[D_n]} \tag{32}$$

$$K_{B,\,id} = \frac{[B]_i[Dis]}{[B]_d[D_n]} \tag{33}$$

The previous equations (Eqs. (30)–(33)) can be rewritten using the characteristic parameters of microemulsions: z and w. Both of them are mole ratios related with the microemulsion geometry. The w ratio is the molar ratio between the disperse phase concentration and the surfactant concentration, and z is the molar ratio between continuous phase concentration and the surfactant concentration (Eqs. (34) and (35)). The w ratio is directly proportional to the droplet radius and z is inversely proportional to the number of microdroplets of the dispersed phase in the microemulsion.

$$w = \frac{[Dis]}{[D_n]} \tag{34}$$

$$z = \frac{[C]}{[D_n]} \tag{35}$$

Hence,

$$K_{A,\,id} = \frac{[A]_i}{[A]_d} w \tag{36}$$

$$K_{B,\,id} = \frac{[B]_i}{[B]_d} w \tag{37}$$

$$K_{A,\,ic} = \frac{[A]_i}{[A]_c} z \tag{38}$$

$$K_{B,ic} = \frac{[B]_i}{[B]_c} z \tag{39}$$

Using Eqs. (28), (29) and (36)–(39), the following equations can be written:

$$[\mathrm{A}]_\mathrm{d} = \frac{[\mathrm{A}]_\mathrm{i} w}{K_{A,id}} \tag{40}$$

$$[\mathrm{B}]_\mathrm{d} = \frac{[\mathrm{B}]_\mathrm{i} w}{K_{B,id}} \tag{41}$$

$$[\mathrm{A}]_\mathrm{c} = \frac{[\mathrm{A}]_\mathrm{i} z}{K_{A,ic}} \tag{42}$$

$$[\mathrm{B}]_\mathrm{c} = \frac{[\mathrm{B}]_\mathrm{i} z}{K_{B,ic}} \tag{43}$$

$$[\mathrm{A}]_\mathrm{T} = [\mathrm{A}]_\mathrm{i} + \frac{[\mathrm{A}]_\mathrm{i} w}{K_{A,id}} + \frac{[\mathrm{A}]_\mathrm{i} z}{K_{A,ic}} \tag{44}$$

$$[\mathrm{B}]_\mathrm{T} = [\mathrm{B}]_\mathrm{i} + \frac{[\mathrm{B}]_\mathrm{i} w}{K_{B,id}} + \frac{[\mathrm{B}]_\mathrm{i} z}{K_{B,ic}} \tag{45}$$

$$[\mathrm{A}]_\mathrm{i} = \frac{[A]_T}{1 + \left(\frac{w}{K_{A,id}}\right) + \left(\frac{z}{K_{A,ic}}\right)} \tag{46}$$

$$[\mathrm{B}]_\mathrm{i} = \frac{[B]_T}{1 + \left(\frac{w}{K_{B,id}}\right) + \left(\frac{z}{K_{B,ic}}\right)} \tag{47}$$

$$[\mathrm{A}]_\mathrm{d} = \left(\frac{w}{K_{A,id}}\right) \frac{[A]_T}{1 + \left(\frac{w}{K_{A,id}}\right) + \left(\frac{z}{K_{A,ic}}\right)} \tag{48}$$

$$[\mathrm{B}]_\mathrm{d} = \left(\frac{w}{K_{B,id}}\right) \frac{[B]_T}{1 + \left(\frac{w}{K_{B,id}}\right) + \left(\frac{z}{K_{B,ic}}\right)} \tag{49}$$

$$[\mathrm{A}]_\mathrm{c} = \left(\frac{z}{K_{A,ic}}\right) \frac{[A]_T}{1 + \left(\frac{w}{K_{A,id}}\right) + \left(\frac{z}{K_{A,ic}}\right)} \tag{50}$$

$$[\mathrm{B}]_\mathrm{c} = \left(\frac{z}{K_{B,ic}}\right) \frac{[B]_T}{1 + \left(\frac{w}{K_{B,id}}\right) + \left(\frac{z}{K_{B,ic}}\right)} \tag{51}$$

Then, using Eqs. (46)–(51), the following expressions are obtained:

$$\begin{aligned} k_2[A_T][B_T] = {} & k_d \left(\frac{w}{K_{A,id}}\right) \frac{[A]_T}{1 + \left(\frac{w}{K_{A,id}}\right) + \left(\frac{z}{K_{A,ic}}\right)} \left(\frac{w}{K_{B,id}}\right) \frac{[B]_T}{1 + \left(\frac{w}{K_{B,id}}\right) + \left(\frac{z}{K_{B,ic}}\right)} + \\ & + k_i \frac{[A]_T}{1 + \left(\frac{w}{K_{A,id}}\right) + \left(\frac{z}{K_{A,ic}}\right)} \frac{[B]_T}{1 + \left(\frac{w}{K_{B,id}}\right) + \left(\frac{z}{K_{B,ic}}\right)} + \\ & + k_c \left(\frac{z}{K_{A,ic}}\right) \frac{[A]_T}{1 + \left(\frac{w}{K_{A,id}}\right) + \left(\frac{z}{K_{A,ic}}\right)} \left(\frac{z}{K_{B,ic}}\right) \frac{[B]_T}{1 + \left(\frac{w}{K_{B,id}}\right) + \left(\frac{z}{K_{B,ic}}\right)} \end{aligned} \tag{52}$$

$$k_2 = k_d\left(\frac{w}{K_{A,id}}\right)\frac{1}{1+\left(\frac{w}{K_{A,id}}\right)+\left(\frac{z}{K_{A,ic}}\right)}\left(\frac{w}{K_{B,id}}\right)\frac{1}{1+\left(\frac{w}{K_{B,id}}\right)+\left(\frac{z}{K_{B,ic}}\right)}+$$
$$+k_i\frac{1}{1+\left(\frac{w}{K_{A,id}}\right)+\left(\frac{z}{K_{A,ic}}\right)}\frac{1}{1+\left(\frac{w}{K_{B,id}}\right)+\left(\frac{z}{K_{B,ic}}\right)}+$$
$$+k_c\left(\frac{z}{K_{A,ic}}\right)\frac{1}{1+\left(\frac{w}{K_{A,id}}\right)+\left(\frac{z}{K_{A,ic}}\right)}\left(\frac{z}{K_{B,ic}}\right)\frac{1}{1+\left(\frac{w}{K_{B,id}}\right)+\left(\frac{z}{K_{B,ic}}\right)} \quad (53)$$

$$k_2 = k_d\left(\frac{w^2}{K_{A,id}K_{B,id}}\right)\frac{1}{1+\left(\frac{w}{K_{A,id}}\right)+\left(\frac{z}{K_{A,ic}}\right)}\frac{1}{1+\left(\frac{w}{K_{B,id}}\right)+\left(\frac{z}{K_{B,ic}}\right)}+$$
$$+k_i\frac{1}{1+\left(\frac{w}{K_{A,id}}\right)+\left(\frac{z}{K_{A,ic}}\right)}\frac{1}{1+\left(\frac{w}{K_{B,id}}\right)+\left(\frac{z}{K_{B,ic}}\right)}+$$
$$+k_c\left(\frac{z^2}{K_{A,ic}K_{B,ic}}\right)\frac{1}{1+\left(\frac{w}{K_{A,id}}\right)+\left(\frac{z}{K_{A,ic}}\right)}\frac{1}{1+\left(\frac{w}{K_{B,id}}\right)+\left(\frac{z}{K_{B,ic}}\right)} \quad (54)$$

$$k_2 = \frac{K_{A,ic}K_{B,ic}w^2k_d}{(K_{A,id}K_{A,ic}+K_{A,ic}w+K_{A,id}z)(K_{B,id}K_{B,ic}+K_{B,ic}w+K_{B,id}z)}+$$
$$+\frac{K_{A,id}K_{A,ic}K_{B,id}K_{B,ic}k_i}{(K_{A,id}K_{A,ic}+K_{A,ic}w+K_{A,id}z)(K_{B,id}K_{B,ic}+K_{B,ic}w+K_{B,id}z)}+$$
$$+\frac{K_{A,id}K_{B,id}z^2k_c}{(K_{A,id}K_{A,ic}+K_{A,ic}w+K_{A,id}z)(K_{B,id}K_{B,ic}+K_{B,ic}w+K_{B,id}z)} \quad (55)$$

$$k_2 = \frac{K_{A,ic}K_{B,ic}w^2k_d+K_{A,id}K_{A,ic}K_{B,id}K_{B,ic}k_i+K_{A,id}K_{B,id}z^2k_c}{(K_{A,id}K_{A,ic}+K_{A,ic}w+K_{A,id}z)(K_{B,id}K_{B,ic}+K_{B,ic}w+K_{B,id}z)} \quad (56)$$

This expression (Eq. (56)) can be simplified considering pseudo-first order conditions, and, of course, taking into account the reagents partitions and the loci of reaction (see **Table 1**).

Finally, to compare the obtained results, as quoted above for micelles model - *vide supra*-, the rate constant values must be corrected taking into account each pseudophase molar volume because the pseudophase model considers that each pseudophase is evenly distributed in the total microemulsion volume [47].

4. Conclusions

The presented model is capable of modelling, as shown in **Table 1**, all the possible circumstances that can occur when the microemulsion is used as a chemical nanoreactor. In all the cases, the adjustment of the experimental data to the model is satisfactory, which shows us that despite its simplicity it presents a great versatility.

We must also indicate that it has not only been applied to micellar systems and microemulsions but also, with satisfactory results, to kinetic processes in other colloidal aggregates such as vesicles [61, 62].

Reaction		**A Partition**			**B Partition**			**Reaction Loci**			**Ref.**
		D	**I**	**C**	**D**	**I**	**C**	**D**	**I**	**C**	
Hydrolysis	Nitrophenyl Acetate (A) + OH^- (B)	■	■	■	■			■			[56]
	Cristal Violet (A) + OH^- (B)	■			■			■			[57]
	Malachite Green (A) + OH^- (B)	■			■			■			[57]
	Sodium nitroprusside (A) + OH^- (B)	■			■			■			[58]
	Carbofuran (A) + OH^- (B)	■	■	■	■			■			[46]
	3-hydroxy-carbofuran (A) + OH^- (B)	■	■	■	■			■			[46]
	3-keto-carbofuran (A) + OH^- (B)	■	■	■	■			■			[46]
Nitrosation	Piperazine (A) + N-Methyl-N-nitroso-p-toluene sulfonamide (B)	■	■			■	■		■		[43]
	N-Methyl-benzyl amine (A) + N-Methyl-N-nitroso-p-toluene sulfonamide (B)		■	■		■	■		■		[43]
	Methyl-ethyl amine (A) + N-Methyl-N-nitroso-p-toluene sulfonamide (B)		■	■		■	■		■		[48]
	Methly-butyl amine (A) + N-Methyl-N-nitroso-p-toluene sulfonamide (B)		■	■		■	■		■		[48]
	Methyl-hexyl amine (A) + N-Methyl-N-nitroso-p-toluene sulfonamide (B)		■	■		■	■		■		[48]
	Methyl-octyl amine (A) + N-Methyl-N-nitroso-p-toluene sulfonamide (B)		■	■		■	■		■		[48]
	Methyl-dodecil amine (A) + N-Methyl-N-nitroso-p-toluene sulfonamide (B)		■	■		■	■		■		[48]
	N-Methyl-benzyl amine (A) + Ethoxy-ethyl nitrite (B)		■	■	■	■	■		■		[44]
	N-Methyl-benzyl amine (A) + Bromo-ethyl nitrite (B)		■	■	■	■	■		■		[44]
	Piperidine (A) + N-Methyl-N-nitroso-p-toluene sulfonamide (B)		■	■		■	■		■		[43]
Nitrosation	Dimiethylamine (A) + N-Methyl-N-nitroso-p-toluene sulfonamide (B)	■	■	■		■	■		■		[43]
	Morphonile (A) + N-Methyl-N-nitroso-p-toluene sulfonamide (B)	■	■	■		■	■		■		[43]
	Pyrrolidine (A) + N-Methyl-N-nitroso-p-toluene sulfonamide (B)	■	■	■		■	■		■		[43]
	Piperazine (A) + Ethoxy-ethyl nitrite (B)	■	■		■	■	■	■	■		[44]
	Piperazine (A) + Bromo-ethyl nitrite (B)	■	■		■	■	■	■	■		[44]
	Morpholine (A) + Ethoxy-ethyl nitrite (B)	■	■	■	■	■	■	■	■		[44]
	Morpholine (A) + Bromo-ethyl nitrite (B)	■	■	■	■	■	■	■	■		[44]
Aminolysis	Sarcosine (A) + Nitrophenyl Acetate (B)	■	■		■	■	■	■	■		[49]
	Piperazine (A) + Nitrophenyl Acetate (B)	■	■		■	■	■	■	■		[49]
	Glycine (A) + Nitrophenyl Acetate (B)	■	■		■	■	■		■		[50]
	N-decyl amine (A) + + Nitrophenyl Acetate (B)		■	■	■	■	■		■		[49]
	N-methyl-benzyl amine (A) + Nitrophenyl Acetate (B)		■	■	■	■	■		■		[51]

Reaction		A Partition			B Partition			Reaction Loci			Ref.
		D	I	C	D	I	C	D	I	C	
	Morpholine (A) + Nitrophenyl Acetate (B)	■	■	■	■	■	■	■	■		[59]
	N-butylamine (A) + Nitrophenyl caprate (B)	■	■	■		■	■		■	■	[60]
Michael addition	Piperazine (A) + N-ethylmaleimide (B)	■	■		■	■	■	■	■		[51]
Solvólisis	Benzoyl chloride (A) + H_2O (B)	■	■	■	■	■		■	■		[53]
	4-Methoxy-benzoyl chloride (A) + H_2O (B)	■	■	■	■	■		■	■		[53]
	Diphenylmethyl chloride (A) + H_2O (B)	■	■	■	■	■		■	■		[53]

D *and* C *correspond to the dispersed or the continuous phase, respectively,* I *denotes the surfactant film. Reactions in w/o microemulsions of AOT/isooctane/water.*

Table 1.
Examples of reactions in microemulsions satisfactorily modelled with the pseudophase model.

Acknowledgements

Financial support from Xunta de Galicia is gratefully acknowledged. Astray G. would like to give his warm thanks to Xunta de Galicia, Consellería de Cultura, Educación e Ordenación Universitaria for the post-doctoral grant B, POS-B/2016/001, which he received from them. Cid A acknowledges the post-doctoral grant SFRH/BD/78849/2011 granted to Requimte and UID/MULTI/04378/2013 granted to Unidade de Ciências Biomoleculares Aplicadas (UCIBIO), both from the Portuguese Foundation for Science and Technology.

This chapter is dedicated to Professor Julio Casado Linarejos (*sit tivi terra levis*) who taught us chemistry and, what really matters, to be better people.

Author details

Antonio Cid[1,2], Aangel Acuña[1,3], Manuel Alonso-Ferrer[1], Gonzalo Astray[1*], Luis García-Río[3], Jesus Simal-Gándara[4] and Juan C. Mejuto[1]

1 Department of Physical Chemistry, Faculty of Science, University of Vigo, Ourense, Spain

2 UCIBIO@REQUIMTE, Chemistry Departamento, Faculdade de Ciências e Tecnologia - Universidade Nova de Lisboa, Caparica, Portugal

3 Departamento de Química Física, Facultad de Química, Universidad de Santiago de Compostela, Santiago de Compostela, Spain

4 Departamento de Química Analítica y Alimentaria, Universidad de Vigo, Ourense, Spain

*Address all correspondence to: gastray@uvigo.es

References

[1] Ruasse MF, Blagoeva IB, Ciri R, García-Río L, Leis JR, Marques A, et al. Organic reaction in micro-organized media: Why and how? Pure and Applied Chemistry. 1997;**69**:1923-1932. DOI: 10.1351/pac199769091923

[2] Menger FM. The structure of micelles. Accounts of Chemical Research. 1979;**12**:111-117. DOI: 10.1021/ar50136a001

[3] Bunton CA, Savelli G. Organic reactivity in aqueous micelles and similar assemblies. Advances in Physical Organic Chemistry. 1986;**22**:213-309. DOI: 10.1016/S0065-3160(08)60169-0

[4] Cordes EH, Dunlap RB. Kinetics of organic reactions in micellar systems. Accounts of Chemical Research. 1969;**2**: 329-337. DOI: 10.1021/ar50023a002

[5] Bunton CA, Romsted LS, Sepulveda L. A quantitative treatment of micellar effects upon deprotonation equilibriums. The Journal of Physical Chemistry. 1980;**84**:2611-2618. DOI: 10.1021/j100457a027

[6] Bunton CA, Romsted LS, Thamvit C. The pseudophase model of micellar catalysis. Addition of cyanide ion it N-alkylpyridinium ions. Journal of the American Chemical Society. 1980;**102**: 3900-3903. DOI: 10.1021/ja00531a036

[7] Bunto CA, Carracsco N, Huang SK, Paik CH, Romsted LS. Reagent distribution and micellar catalysis of carbocation reactions. Journal of the American Chemical Society. 1978;**100**: 5420-5425. DOI: 10.1021/ja00485a028

[8] Bravo C, Herves P, Leis JR, Peña ME. Basic hydrolysis of N-[N′methyl-N′-nitroso(aminomethyl)]benzamide in aqueous and micellar media. Journal of Colloid and Interface Science. 1992;**153**: 529-536. DOI: 10.1016/0021-9797(92) 90343-K

[9] Vera S, Rodenas E. Inhibiton effect of cationic micelles on the basic hydrolysis of aromatic esters. Tetrahedron. 1986; **42**:143-149. DOI: 10.1016/S0040-4020 (01)87411-1

[10] Bravo C, Leis JR, Peña ME. Effect of alcohols on catalysis by dodecyl sulfate micelles. The Journal of Physical Chemistry. 1992;**96**:1957-1961. DOI: 10.1021/j100183a077

[11] Bunton CA, Romsted LS, Savelli G. Tests of the pseudophase model of micellar catalysis: Its partial failure. Journal of the American Chemical Society. 1979;**101**:1253-1259. DOI: 10.1021/ja00499a034

[12] Al-Lohedan H, Bunton CA, Romsted LS. Micellar effects upon the reaction of betaine esters with hydroxide ion. The Journal of Physical Chemistry. 1981;**85**: 2123-2129. DOI: 10.1021/j150614a034

[13] Bunton CA, Hamed FH, Romsted LS. Quantitative treatment of reaction rates in functional micelles and comicelles. The Journal of Physical Chemistry. 1982;**86**:2103-2108. DOI: 10.1021/j100208a040

[14] Bunton CA, Romsted LS, Smith HJ. Quantitative treatment of micellar catalysis of reactions involving hydrogen ions. The Journal of Organic Chemistry. 1978;**43**:4299-4303. DOI: 10.1021/jo00416a010

[15] Bunton CA, Gan LH, Moffatt JR, Romensted LS, Savelli G. Reactions in micelles of cetyltrimethylammonium hydroxide. Test of the pseudophase model for kinetics. The Journal of Physical Chemistry. 1981;**85**:4118-4125. DOI: 10.1021/j150626a033

[16] Bravo-Díaz C, Romero-Nieto MA, Gonzalez-Romero E. Micellar-promoted homolytic dediazoniation of

p-nitrobenzenediazonium tetrafluoroborate. Langmuir. 2000;**16**: 42-48. DOI: 10.1021/la9901682

[17] Bravo C, Hervés P, Leis JR, Peña ME. Micellar effects in the acid denitrosation of N-nitroso-N-methyl-p-toluenesulfonamide. The Journal of Physical Chemistry. 1990;**94**:8816-8820. DOI: 10.1021/j100388a014

[18] Costas-Costas U, Bravo-Díaz C, Gonzalez-Romero E. Micellar effects on the reaction between an arenediazonium salt and 6-o-octanoyl-l-ascorbic acid. Kinetics and mechanism of the reaction. Langmuir. 2004;**20**: 1631-1638. DOI: 10.1021/la036142z

[19] Ródenas E, Valiente M, Villafruela MS. Different theoretical approaches for the study of the mixed tetraethylene glycol mono-n-dodecyl ether/ hexadecyltrimethylammonium bromide micelles. The Journal of Physical Chemistry. B. 1999;**103**:4549-4554. DOI: 10.1021/jp981871m

[20] Ortega F, Ródenas E. Micellar effects upon the reaction of low-spin diimine-iron(II) complexes with hydroxide and cyanide ions. The Journal of Physical Chemistry. 1986; **90**:2408-2413. DOI: 10.1021/ j100402a031

[21] García-Río L, Leis JR, Mejuto JC, Perez-Juste J. Hydrolysis of N-methyl-N-nitroso-p-toluenesulphonamide in micellar media. Journal of Physical Organic Chemistry. 1998;**11**:584-588. DOI: 10.1002/(SICI) 1099-1395(199808/09)11:8/9< 584::AID-POC59>3.0.CO;2-F

[22] García-Rio L, Hervés P, Leis JR, Mejuto JC, Rodríguez-Dafonte P. Reactive micelles: Nitroso group transfer from N-methyl-N-nitroso-p-toluenesulfonamide to amphiphilic amines. Journal of Physical Organic Chemistry. 2004;**17**:1067-1072. DOI: 10.1002/poc.827

[23] Astray G, Cid A, Manso JA, Mejuto JC, Moldes O, Morales J. Influence of anionic and nonionic micelles upon hydrolysis of 3-hydroxy-carbofuran. International Journal of Chemical Kinetics. 2011;**43**:402-408. DOI: 10.1002/kin.20563

[24] Arias M, García-Río L, Mejuto JC, Rodríguez-Dafonte P, Simal-Gándara J. Influence of micelles on the basic degradation of carbofuran. Journal of Agricultural and Food Chemistry. 2005;**53**:7172-7178. DOI: 10.1021/ jf0505574

[25] García-Rio L, Mejuto JC, Pérez-Lorenzo M, Rodríguez-Dafonte P. NO transfer reactions between N-methyl-N-nitroso-p-toluene sulfonamide and N-alkylamines in CTACl micellar aggregates. Progress in Reaction Kinetics and Mechanism. 2006;**31**:129-138. DOI: 10.3184=146867806X197115

[26] Astray G, Cid A, Manso JA, Mejuto JC, Moldes OA, Morales J. Alkaline fading of triarylmethyl carbocations in self-assembly microheterogeneous media. Progress in Reaction Kinetics and Mechanism. 2011;**36**:139-165. DOI: 10.3184/146867811X12984793755693

[27] Shrivastava A, Singh AK, Sachdev N, Shrivastava DR, Katre Y, Singh SP, et al. Micelle catalyzed oxidative degradation of norfloxacin by chloramine-T. Journal of Molecular Catalysis A. 2012;**361–362**:1-11. DOI: 10.1016/j.molcata.2012.04.004

[28] Morales J, Moldes OA, Cid A, Astray G, Mejuto JC. Cleavage of carbofuran and carbofuran-derivatives in micellar aggregates. Progress in Reaction Kinetics and Mechanism. 2015;**40**: 105-118. DOI: 10.3184/ 146867815X14259195615547

[29] Bunton CA, Nome F, Quina FH, Romsted LS. Ion binding and reactivity at charged aqueous interfaces. Accounts

of Chemical Research. 1991;**24**:357-364. DOI: 10.1021/ar00012a001

[30] Amado S, García-Rio L, Leis JR, Rios A. Reactivity of anions with organic substrates bound to sodium dodecyl sulfate micelles: A Poisson–Boltzmann/pseudophase approach. Langmuir. 1997;**13**:687-692. DOI: 10.1021/la960749g

[31] Fernández I, García-Río L, Hervés P, Mejuto JC, Pérez-Juste J, Rodríguez-Dafonte P. β-Cyclodextrin–micelle mixed systems as a reaction medium. Denitrosation of N-methyl-N-nitroso-p-toluenesulfonamide. Journal of Physical Organic Chemistry. 2000;**13**:664-669. DOI: 10.1002/1099-1395(200010)13:10<664::AID-POC264>3.0.CO;2-U

[32] Astray G, Cid A, García-Río L, Lodeiro C, Mejuto JC, Moldes O, et al. Cyclodextrin-surfactant mixed systems as reaction media. Progress in Reaction Kinetics and Mechanism. 2010;**35**:105-129. DOI: 10.3184/146867810X12686717520194

[33] García-Río L, Leis JR, Mejuto JC, Pérez-Juste J. Basic hydrolysis of m-nitrophenyl acetate in micellar media containing β-cyclodextrins. The Journal of Physical Chemistry. B. 1998;**102**:4581-4587. DOI: 10.1021/jp980432k

[34] Alvarez AR, García-Rio L, Hervés P, Leis JR, Mejuto JC, Perez-Juste J. Basic hydrolysis of substituted nitrophenyl acetates in β-cyclodextrin/surfactant mixed systems. Evidence of free cyclodextrin in equilibrium with micellized surfactant. Langmuir. 1999;**15**:8368-8375. DOI: 10.1021/la981392e

[35] García-Río L, Leis Jr, Mejuto JC, Navarro-Vázquez A, Pérez-Juste J, Rodríguez-Dafonte P. Basic hydrolysis of crystal violet in β-cyclodextrin/surfactant mixed systems. Langmuir. 2004;**20**:606-613. DOI: 10.1021/la035477d

[36] Dorrego AB, García-Rio L, Hervés P, Leis JR, Mejuto JC, Pérez-Juste J. Micellization versus cyclodextrin-Surfactant complexation. Angewandte Chemie, International Edition. 2000;**39**:2945-2948. DOI: 10.1002/1521-3773(20000818)39:16<2945::AID-ANIE2945>3.0.CO;2-6

[37] Dorrego B, García-Rio L, Hervés P, Leis JR, Mejuto JC, Pérez-Juste J. Changes in the fraction of uncomplexed cyclodextrin in equilibrium with the micellar system as a result of balance between micellization and cyclodextrin-surfactant complexation. cationic alkylammonium surfactants. The Journal of Physical Chemistry. B. 2001;**105**:4912-4920. DOI: 10.1021/jp0035232

[38] Astray G, García-Río L, Lodeiro C, Mejuto JC, Moldes O, Morales J, et al. Influence of colloid suspensions of humic acids on the alkaline hydrolysis of N-methyl-N-nitroso-p-toluene sulfonamide. International Journal of Chemical Kinetics. 2010;**42**:316-322. DOI: 10.1002/kin.20481

[39] Morales J, Manso JA, Cid A, Mejuto JC. Stability study of iprodione in alkaline media in the presence of humic acids. Chemosphere. 2013;**92**:1536-1541. DOI: 10.1016/j.chemosphere.2013.04.020

[40] Morales J, Manso JA, Cid A, Mejuto JC. Degradation of carbofuran and carbofuran-derivatives in presence of humic substances under basic conditions. Chemosphere. 2012;**89**:1267-1271. DOI: 10.1016/j.chemosphere.2012.05.018

[41] Morales J, Cid A, Mejuto JC. Alkaline hydrolysis of vinclozolin: Effect of humic acids aggregates in water. Journal of Molecular Catalysis A. 2015;**401**:13-17. DOI: 10.1016/j.molcata.2015.02.017

[42] Arias-Estevez M, Astray G, Cid A, Fernández-Gándara D, García-Río L,

Mejuto JC. Influence of colloid suspensions of humic acids upon the alkaline fading of carbocations. Journal of Physical Organic Chemistry. 2007;**21**: 555-560. DOI: 10.1002/poc.1317

[43] García-Río L, Leis JR, Peña ME, Iglesias E. Transfer of the nitroso group in water/AOT/isooctane microemulsions: Intrinsic and apparent reactivity. The Journal of Physical Chemistry. 1993;**97**:3437-3442. DOI: 10.1021/j100115a057

[44] García-Río L, Leis JR, Mejuto JC. Pseudophase approach to reactivity in microemulsions: Quantitative explanation of the kinetics of the nitrosation of amines by alkyl nitrites in AOT/isooctane/water microemulsions. The Journal of Physical Chemistry. 1996;**100**:10981-10988. DOI: 10.1021/jp953264u

[45] Astray G, Cid A, García-Río L, Hervella P, Mejuto JC, Pérez-Lorenzo M. Organic reactivity in AOT-stabilized microemulsions. Progress in Reaction Kinetics and Mechanism. 2008;**33**:81-97. DOI: 10.3184/146867807X273173

[46] Morales J, Manso JA, Cid A, Lodeiro C, Mejuto JC. Degradation of carbofuran derivatives in restricted water environments: Basic hydrolysis in AOT-based microemulsions. Journal of Colloid and Interface Science. 2012;**372**: 113-120. DOI: 10.1016/j.jcis.2012.01.022

[47] García-Río L, Leis JR, Mejuto JC, Perez-Lorenzo M. Microemulsions as microreactors in physical organic chemistry. Pure and Applied Chemistry. 2007;**79**:1111-1123. DOI: 10.1351/pac200779061111

[48] García-Río L, Hervés P, Mejuto JC, Pérez-Juste J, Rodríguez-Dafonte P. Pseudophase approach to reactivity in microemulsions: Quantitative explanation of the kinetics of the nitroso group transfer reactions between N-methyl-N-nitroso-p-toluenesulfonamide and secondary alkylamines in water/AOT/isooctane microemulsions. Industrial and Engineering Chemistry Research. 2003; **42**:5450-5456. DOI: 10.1021/ie0208523

[49] García-Rio L, Mejuto JC, Pérez-Lorenzo M. Microheterogeneous solvation for aminolysis reactions in AOT-based water-in-oil microemulsions. Chemistry - A European Journal. 2005;**11**:4361-4373. DOI: 10.1002/chem.200401067

[50] García-Río L, Mejuto JC, Pérez-Lorenzo M. Aminolysis reactions by glycine in AOT-based water-in-oil microemulsions. Colloids and Surfaces, A: Physicochemical and Engineering Aspects. 2005;**270–271**:115-123. DOI: 10.1016/j.colsurfa.2005.05.048

[51] Fernández E, García-Río L, Leis JR, Mejuto JC, Pérez-Lorenzo M. Michael addition and ester aminolysis in w/o AOT-based microemulsions. New Journal of Chemistry. 2005;**29**: 1594-1600. DOI: 10.1039/b507190a

[52] García-Río L, Hervés P, Mejuto JC, Pérez-Juste J, Rodríguez-Dafonte P. Pseudophase approach to the transfer of the nitroso group in water/AOT/SDS/isooctane quaternary microemulsions. Langmuir. 2000;**16**:9716-9721. DOI: 10.1021/la000523k

[53] García-Río L, Leis JR, Mejuto JC. Solvolysis of benzoyl halides in AOT/isooctane/water microemulsions. Influence of the Leaving Group. Langmuir. 2003;**19**:3190-3197. DOI: 10.1021/la026753b

[54] García-Río L, Hervés P, Mejuto JC, Rodríguez-Dafonte P. Nitrosation reactions in water/AOT/xylene microemulsions. Industrial and Engineering Chemistry Research. 2006; **45**:600-606. DOI: 10.1021/ie050925t

[55] García-Rio L, Mejuto JC, Pérez-Lorenzo M. Simultaneous effect of

microemulsions and phase-transfer agents on aminolysis reactions. The Journal of Physical Chemistry. B. 2007; **111**:11149-11156. DOI: 10.1021/jp0743323

[56] García-Río L, Mejuto JC, Pérez-Lorenzo M. Modification of reactivity by changing microemulsion composition. Basic hydrolysis of nitrophenyl acetate in AOT/isooctane/water systems. New Journal of Chemistry. 2004;**28**:988-995. DOI: 10.1039/b401226g

[57] Leis JR, Mejuto JC, Peña ME. Comparison between the kinetics of the alkaline fading of carbocation dyes in water/sodium bis(2-ethylhexyl) sulfosuccinate/isooctane microemulsions and in homogeneous media. Langmuir. 1993;**9**:889-893. DOI: 10.1021/la00028a003

[58] García-Río L, Hervés P, Leis JR, Mejuto JC, Pérez-Juste J. Determination of the hydrolysis rate of AOT in AOT/isooctane/water microemulsions using sodium nitroprusside as chemical probe. Journal of Physical Organic Chemistry. 2002;**15**:576-581. DOI: 10.1002/poc.513

[59] García-Río L, Mejuto JC, Pérez-Lorenzo M. Ester aminolysis by morpholine in AOT-based water-in-oil microemulsions. Journal of Colloid and Interface Science. 2006;**301**:624-630. DOI: 10.1016/j.jcis.2006.05.037

[60] García-Río L, Mejuto JC, Pérez-Lorenzo M. First evidence of simultaneous different kinetic behaviors at the interface and the continuous medium of w/o microemulsions. The Journal of Physical Chemistry. B. 2006; **110**:812-819. DOI: 10.1021/jp055270o

[61] Hervés P, Leis JR, Mejuto JC, Pérez-Juste J. Kinetic studies on the acid and alkaline hydrolysis of N-methyl-N-nitroso-p-toluenesulfonamide in dioctadecyldimethylammonium chloride vesicles. Langmuir. 1997;**13**: 6633-6637. DOI: 10.1021/la9705975

[62] García-Río L, Hervés P, Mejuto JC, Pérez-Juste J, Rodríguez-Dafonte P. Comparative study of nitroso group transfer in colloidal aggregates: Micelles, vesicles and microemulsions. New Journal of Chemistry. 2003;**27**:372-380. DOI: 10.1039/b209539d

Chapter 2

Nitrosation of Amines in AOT-Based Microemulsions

Pedro Rodríguez-Dafonte

Abstract

This chapter is a review of the kinetics of nitrosation of secondary amines by N-methyl-N-nitroso-p-toluenesulfonamide (MNTS) in AOT-based microemulsions. Three regions can be distinguished in these colloids: the internal aqueous nanocore, the micellar interface and the external organic phase. The amines were chosen on the basis of their degrees of solubility resulting in a different distribution. The MNTS has a very low degree of solubility in water and the nitrosation reactions take place at the interface of the aggregates. The polarity changes at the interface have very important effects on the chemical reactivity. This kinetic study compares the results obtained in AOT microemulsions where the polarity at the interface can be tuned by adding a cosurfactant or by changing the continuous medium.

Keywords: AOT, MNTS, nitrosation, kinetics, amines

1. Introduction

Microemulsions are macroscopically homogeneous, thermodynamically stable systems constituted by two immiscible solvents in the presence of a surfactant [1]. Water in oil microemulsions consist of nanodroplest of water dispersed in a nonpolar solvent stabilized by a surfactant monolayer. Sodium bis(2-ethylhexyl) sulfosuccinate (AOT) is the most widely used surfactant for preparing microemulsions. The simplest microstructure AOT-based systems are that of spherical water droplets of colloidal dimensions and possess a small degree of polydispersity [2]. The size can be accurately controlled by the water content or, that is, the molar ratio of water to AOT ($W = [H_2O]/[AOT]$).

Transparent water-in-oil dispersions were reported for the first time in 1943 [3]. Hoar and Schulman found that oil can be dissolved in bulk water or water in bulk oil in presence of a soap to produce a homogeneous solution. The oil-surfactant-water water systems reported were transparent, electrically non-conducting dispersions, in which the oil was the continuous phase. Dilution of these systems with excess water inverted them to oil-in-water systems. The key factor was the oil/surfactant or water/surfactant ratio (W). Later, Schulman et al. [4], by electron microscopy technique established that such dispersions consist of uniform spherical droplets of either oil or water dispersed in the appropriate continuous phase and were defined as microemulsions.

The first use of AOT as appropriate surfactant in the preparation of ternary systems was published in 1970 [5]. The authors studied the phase equilibria diagram of the following systems: p-xylene/AOT/water, caprylic acid/AOT/water, and n/decanol/AOT/water. However, isooctane is actually the most widely oil used in the preparation of AOT-based microemulsions, due to the fact that this solvent show a large region of isotropic transparent solutions, thermodynamically stable. The complete phase equilibria diagram was reported in 1980 [6, 7]. **Figure 1** shows the structure of the isooctane/AOT/water microemulsion.

Among the many applications of these systems, the use of the water pool as a nanoreactor for the preparation of nanoparticle stands out. The nanoparticles size is mainly controlled by the microemulsion size. For instance, ZnS nanoparticles were synthetized and characterized in n-heptane/AOT/water microemulsions [8, 9] and the nanoparticle size can be controlled by the W parameter. Indium-tin oxide (In_2O_3/SnO_2) nanoparticles were prepared in isooctane/AOT/water system [10]. The molecular structure of AOT favors the interface curved on the water core. The result is that the monodispersed ultrafine indium-tin oxide particles show better physicochemical properties prepared in microemulsion than in bulk precipitation method. In the same system barium chromate nanostructures (linear chains, rectangular superlattices and long filaments) as a function of reactant molar ratio were prepared [11]. Another example is the synthesis of crystalline nanoparticles of three different molecule-based magnetic materials, cobalt hexacyanoferrate, cobalt pentacyanonitrosylferrate, and chromium hexacyanochromate, by coprecipitation reactions involving mixtures of water-in-oil microemulsions [12].

However, if the objective of the research is the study of microemulsions as chemical nanoreactors, it is necessary to study the details of the processes that take place inside them. Our research group has focused, for more than 20 years, on the study of reaction mechanisms in a wide variety of colloidal systems. From a kinetic point of view, microemulsions are more versatile than other colloids because they provide both organic and aqueous environments. Then, it's possible to simultaneously dissolve both hydrophobic and hydrophilic reagents, each compound being distributed among oil, water, and surfactant film in accordance with its physicochemical nature. In the case of water in oil microemulsions, the properties of the confined water are very different from those of the bulk water due to intermolecular interactions at the micellar interface and to the geometric size constraints of the environment [13]. This characteristic makes possible the use of

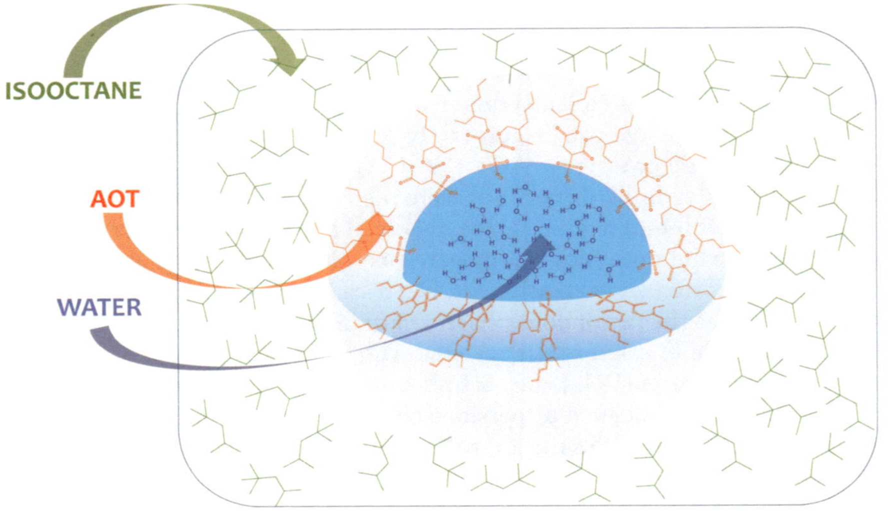

Figure 1.
Schematic illustration of a isooctane/AOT/water microemulsion.

microemulsions as biomimetic models for compartmentalization [14]. Most of these initial kinetic studies have not been able to explain the observed kinetic behavior quantitatively.

In 1993, in our laboratory, García-Río et al. [15] reported the use of the pseudophase formalism as modified for microemulsions. The reaction studied (see **Figure 2**) was the nitrosation of secondary amines by N-methyl-N-nitroso-p-toluenesulfonamide (MNTS) in isooctane/AOT/water microemulsions. The amines studied were piperazine, N-methylbenzylamine, piperidine, dimethylamine, morpholine, pyrrolidine, and diisopropylamine. In that work, the validity of the model for reactions taking place at interface between the water droplet and the isooctane was confirmed. The pseudo-phase model continued to be applied successfully in many reactions carried out in this type of microemulsions. For instance, for nitroso group transfer from 2-ethoxyethyl and 2-bromoethyl nitrite to the secondary amines piperazine, N-methylbenzylamine, and morpholine in isooctane/AOT/water microemulsions [16]. The diverse kinetic behavior was explained quantitatively based on the model considering the distribution of the amine among the aqueous and isooctane phases and their mutual interface; the reaction itself always takes place at the interface. Among all these reactions, we are especially interested in the MNTS reaction with piperazine (PIP), N-methylbenzylamine (NMBA), and morpholine (MOR) (see **Figure 3**). The reason is that these three species represent three different solubilities in water in oil microemulsions. Piperazine is practically insoluble in isooctane, N-methylbenzylamine is poorly soluble in water, and morpholine has considerable solubility in both water and isooctane.

In the kinetic studies carried out in microemulsions, the relative reactivities of the amines are discussed in comparison with those observed in bulk water. The comparison requires knowledge of the molar reaction volume at the interface. An estimation of this volume was made and yields a value of 0.37 M^{-1} for the molar volume of AOT in isooctane/AOT/water microemulsions. The nitroso transfer reactions are 58 times slower at the interface of the microemulsion, which can be attributed to the lower polarity of the interfacial region [17].

In this chapter our objective is to show how the reactivity of AOT-based microemulsions can change due to changes in the continuous medium or the introduction of an anionic cosurfactant at the interface. In colloidal systems, with applications in cosmetics and cleaning products, it is usual to use two surfactants with very different properties. In the case of micellar aggregation, mixtures of surfactants often work better than a single surfactant. For example, in detergents, anionic surfactants are included to maximize solubilization, and non-ionic surfactants to

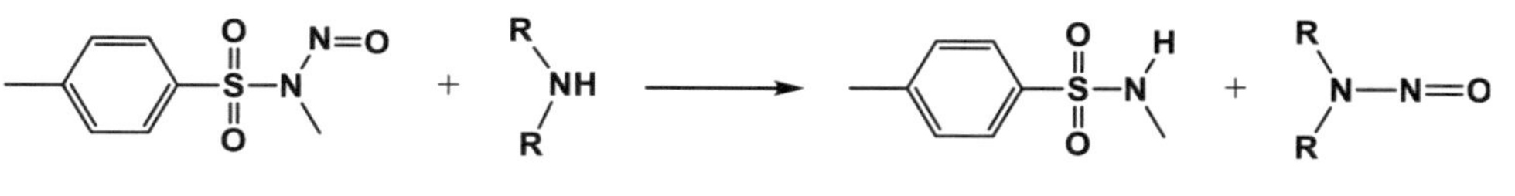

Figure 2.
Nitrosation reaction of secondary amines by N-methyl-N-nitroso-p-toluenesulfonamide (MNTS).

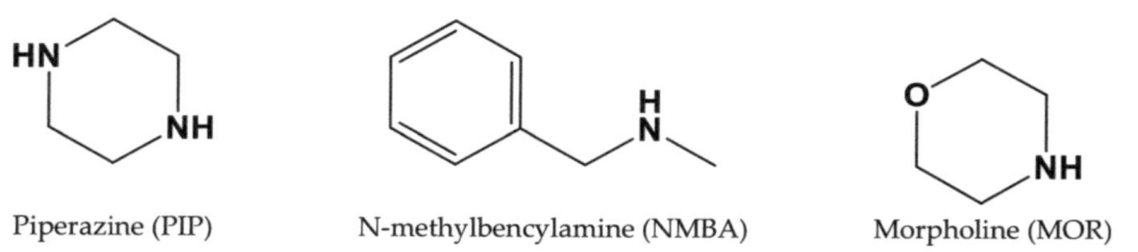

Figure 3.
Structure of the amines.

increase water hardness tolerance. Mixed microemulsions are formed generally by cationic surfactants and long-chain alcohols. Structural and physiochemical characterization of these systems have been published. For example, the position of the alcohol between the interfacial and the bulk oil phase was determined by different experimental techniques [18–21]. Kinetic studies in microemulsions with two surfactants were also carried out in our laboratory. For instance, García-Rio and Hervella reported the nitrosation of piperazine (PIP) and N-methylbenzylamine (NMBA) by N-methyl-N-nitroso-p-toluenesulfonamide (MNTS) in microemulsions of isooctane/tetradecyltrimethylammonium bromide (TTABr)/alcohol/water, varying nature of the alcohols: 1-pentanol, 1-hexanol, 1-heptanol, 1-octanol and 1-decanol [22]. The main conclusion of this article is that the effect of the alcohols is the increase in volume of the interface with the consequent dilution of the reactants. The values of the rate constants obtained in TTABr/alcohol microemulsions were always lower than those obtained in AOT microemulsions. The incorporation of the alcohol into the interface of the microemulsion displaces the water molecules and increases its hydrophobic character.

In this chapter, we present the comparative review of the nitrosation of secondary amines by MNTS nitrosation in two clearly differentiated systems:

1. Microemulsions with AOT and SDS as surfactants [23]. The presence of SDS as a cosurfactant causes a sharp decrease in the electrical percolating temperature, that is, the surfactant film becomes less rigid favoring the transport of matter. The amines (piperazine, N-methylbenzylamine and morpholine) were chosen based on their degrees of solubility in the different components of the microemulsions.

2. AOT-based microemulsions with ortho-, meta- or para-xylene as continuous medium [24]. The use of an aromatic molecule as the continuous medium of a microemulsion modifies the reactivity. In addition, this type of microemulsions has the characteristic of being non-percolative. The amines were chosen based on their different degrees of solubility in the different components of the microemulsion and on the fact that numerous kinetic studies had already been carried out in “normal” microemulsions.

2. Percolation temperature

The first step, to characterize these systems is to analyze the effect of the co-surfactant or the oil on the microstructure of the aggregates. Microemulsions are dynamic structures and electrical conductivity measurements can be a useful method for characterizing aggregate interactions [25]. Microemulsions have a low electrical conductivity (0.01–0.1 $\mu S\ cm^{-1}$), but higher than that of pure isooctane ($<1 \times 10^{-3}\ \mu S\ cm^{-1}$) because they are systems that contain charges. Increasing the temperature, the electrical conductivity of these systems increases gradually, at a temperature from which there is a marked increase in the variation of the electrical conductivity with temperature ($>100\ \mu S\ cm^{-1}$). This singularity is known as electrical percolation, and the temperature at which this occurs is known as the threshold of percolation or the temperature of percolation. The values of the threshold of percolation, in a certain microemulsion, can be modified by small quantities of additives. The electrical percolation phenomenon is explained assuming a mechanism involving the formation of channels through which mass is exchanged between disperse water droplets in the continuous phases. An effective collision between two water nanodroplets is necessary to allow the droplets to fuse together.

Then, mass transfer between the water droplets must take place to allow charge to be conducted and the droplets to separate by fission.

AOT-based microemulsions, thermodynamically stable over a wide range of water proportions, are the most extensively examined. The size of these aggregates can be controlled via the parameter W (W = $[H_2O]/[AOT]$) as the droplet radius is directly related to W, radius = 1.5 W [26]. In the case of ternary isooctane/water/AOT microemulsions, the temperature of percolation decreases by the presence of formamides and urea [27]. However, the presence of amines increases the threshold of percolation [28]. An important effect is observed when adding SDS to water/AOT/isooctane. An increase in the molar ratio ρ = [SDS]/AOT], in isooctante/AOT/SDS/water/microemulsion corresponds with a strong decrease in the percolation threshold for these quaternary microemulsions (see **Figure 4**). The difference between the percolation threshold of a standard AOT-based microemulsion and the quaternary system is more than 40°C. A similar effect was found for the rate constant of matter exchange between droplets. In the case of heptane/AOT/SDS/water system, the value of the rate constant is $1.3 \times 10^7\ M^{-1}\ s^{-1}$ for ρ = 0 (ternary microemulsion) and $2.2 \times 10^7\ M^{-1}\ s^{-1}$ for ρ = 0.1 (quaternary microemulsion) [29]. The explanation was that SDS tends to locate at the water-hydrocarbon interface, thus causing a disruption of the surfactant head groups at the interface, which, in turn, promotes a more rapid exchange of pool components on pool encounters.

Maximum water solubilization capacity of the quaternary system was studied by adding an appropriate volume of water to samples containing known amounts of AOT, SDS and isooctane under continuous stirring until permanent turbidity was observed. An increase in W increases the overlap volume between two droplets as they approach each other to form a cluster, thereby increasing the interaction potential and facilitating percolation. While droplet attraction plays a prominent role in percolation, the nature of the interface is also relevant to the process because mass transfer is impossible unless the interface is ruptured in some way. In the case of isooctane/AOT/water microemulsions the presence in the aqueous phase of a moderate concentration of organic additives decreased the solubilization capacity and led to phase separation at a W value between 47 and 42 for thiourea compounds [30]. If the additive is an electrolyte ($NaClO_4$, NH_4Cl, NaBr and Na_2SO_4) the water

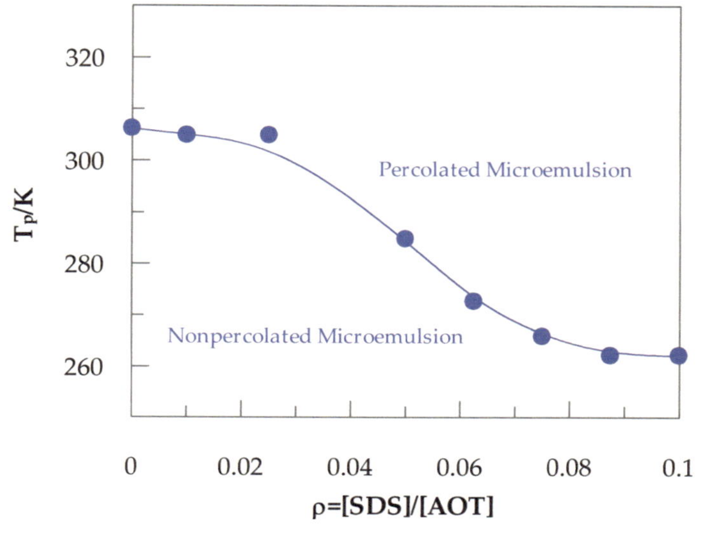

Figure 4.
Changes in percolation temperatures of water/AOT/SDS/isooctane microemulsions for different ρ values (ρ = [SDS]/AOT]). The solid line is a guide for the eye. W = 22, [AOT] = 0.5 M.

solubilization capacity ($W_{breakage}$) decreases to W = 36–30. In our quaternary system, **Figure 5**, shows how $W_{breakage}$ decreases by increasing the amount of SDS as cosurfactant. SDS modifies the natural negative curvature of the AOT by increasing the rigidity of its film and hindering its distortion, thereby decreasing interactions among droplets.

In this chapter, we also examined the xylene/AOT/water system, using o-, m-, and p-xylene. Percolation temperature of water in oil microemulsions is known to depend on the properties of the bulk solvent [31]. By using n-decane, n-octane, and n-heptane as continuous media, it was found that the percolation temperature to increase with decreasing length of the alkyl chain in the hydrocarbon. The strong attractive interaction between water nanodroplets was assumed to be the major

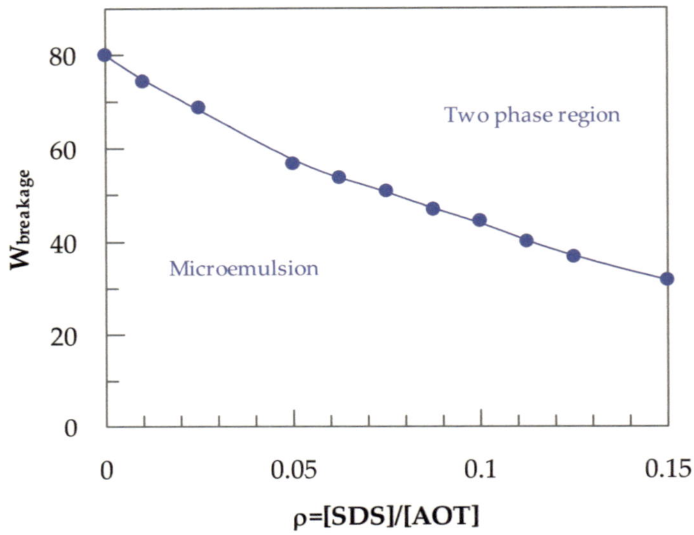

Figure 5.
Influence of ρ (ρ = [SDS]/AOT]) in the water solubilization capacity ($W_{breakage}$) of isooctane/AOT/SDS/water microemulsions. The solid line is a guide for the eye. W = 22, [AOT] = 0.5 M.

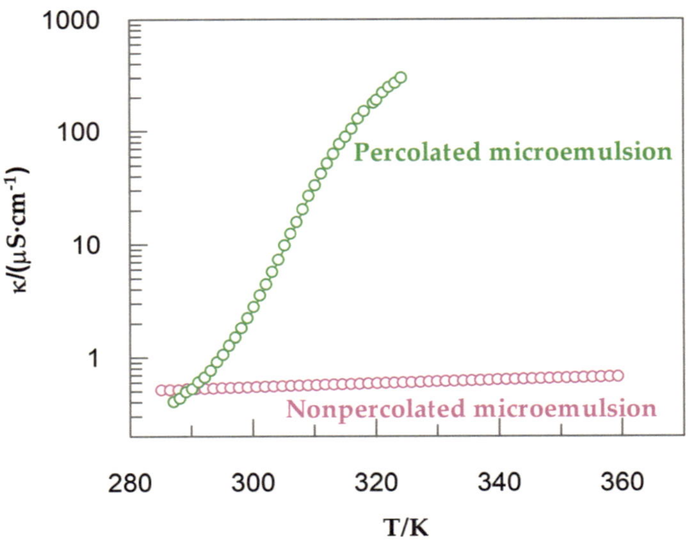

Figure 6.
Conductivity vs. temperature plot for an isooctane/AOT/water microemulsion (○) and an o-xylene/AOT/water microemulsion (○). [AOT] = 0.5 M, W = 12.

factor limiting microemulsion stability. An increase in molar mass of the hydrocarbon oil was found to decrease the density of the continuous phase and facilitate access of the interacting droplets for mass transfer. Microemulsions with xylenes as continuum medium are non-percolative and **Figure 6** shows the variation of conductivity with temperature for o-xylene/AOT/water and isooctane/AOT/SDS/water microemulsions.

The goal of this chapter is to conduct a comparative kinetic study on percolative and nonpercolative microemulsions. The difference in composition between the systems should result in differences in structure and physical properties as the ease with which the oil can penetrate the surfactant layer will vary markedly with the steric hindrance it encounters.

3. Nitrosation in isooctane/AOT/SDS/water microemulsions

The kinetics, at 25°C, of the transfer of the nitroso group from MNTS to secondary amines (PIP, NMBA and MOR) was studied using a wide variety of quaternary microemulsions as reaction media. MNTS reacts with secondary amines by transnitrosation to give carcinogenic *N*-nitrosamines [32] and, for this reason, the chemistry of MNTS has received increasing interest. The reactivity, in water, between MNTS and amines has been reported in our laboratory [33]. Kinetic studies were carried out in AOT-based microemulsions and micellar media [34–38]. The rate constants obtained for the reactions in microemulsions were in all cases like those in water, with first-order terms in MNTS and total amine concentration. The pseudophase model allows us to provide a quantitative explanation of the experimental results. The model considers the distribution of the amine between the oil, the surfactant film and the water nanodroplet. The MNTS has a very low degree of solubility in water and the nitrosation reaction always takes place at the interface. The presence of SDS changes the properties at the interface of the microemulsion, and thus the reactivity.

The influence of the composition of the quaternary system on the rate of transnitrosation of PIP, NMBA and MOR by MNTS was studied in a series of experiments in which the nanodroplet size varied from series to series over the range W = 7.4–25.8. The total surfactant concentration was varied between 0.443 and 0.544. The mole ratio ρ (ρ = [SDS]/[AOT) was varied from series to series between 0 and 0.15.

The reaction kinetics was followed by monitoring the decrease in absorbance, at the appropriate wavelength, due to MNTS consumption. The concentration of MNTS was always much lower than the concentration of amine. The absorbance-time data of all kinetic experiments were fitted by first-order integrated equations, and the values of the pseudo-first-order rate constants, $\boldsymbol{k_o}$, were reproducible to within 5%.

3.1 Reaction with piperazine (PIP)

The influence of the structure of the quaternary system on the rate of transnitrosation of PIP by MNTS was studied in a series of experiments in which total concentration of pierazine was 0.05 M. **Figure** 7 shows that the value of pseudo-first-order rate constants (k_o) increases by increasing total surfactant concentration. However, k_o decreases by increasing the nanodroplet size (W).

Kinetic studies of reactions AOT-based microemulsions can be analyzed in terms of reactivity only if local reagent concentrations and intrinsic rate constants in the various microphases of these organized media can be obtained from the overall apparent rate data. To apply the pseudophase formalism, we must consider the microemulsion formed by three strongly differentiated pseudophases: a continuous medium formed fundamentally by organic solvent (oil) an aqueous pseudophase (water) and an interface formed fundamentally by the surfactants (interface). Because of the very low water solubility of MNTS it will partition between the continuous medium and surfactant film, where the reaction is taking place. However, PIP is distributed between water and the surfactant film. **Figure 8** shows schematically the distribution of PIP and MNTS between the three phases of the microemulsion.

The partition coefficients that define the distribution of the reagents among the three pseudophases are defined in terms of their mole per mole concentrations in the pseudophases:

$$K_1 = \frac{[Amine]_{interface}}{[Amine]_{water}} W \quad (1)$$

$$K_4 = \frac{[MNTS]_{interface}}{[MNTS]_{oil}} Z \quad (2)$$

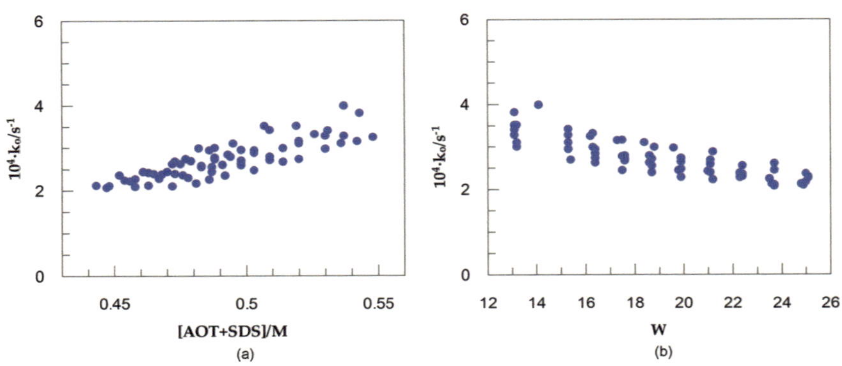

Figure 7.
Left: influence of the concentration of surfactant upon k_o, at 25°C, for the nitrosation of piperazine by MNTS. Right: influence of the amount of water (W, W = $[H_2O]/([AOT] + [SDS])$) upon k_o, at 25°C, for the nitrosation of piperazine by MNTS.

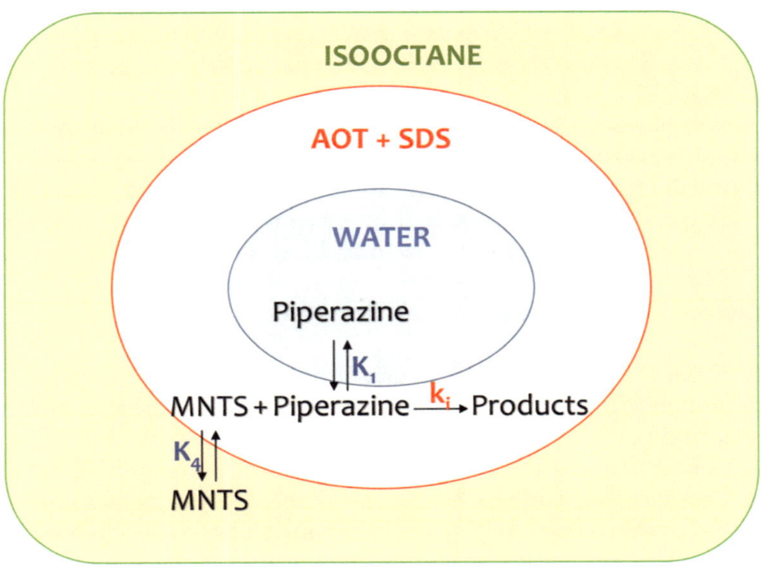

Figure 8.
Pseudophase model for the nitrosation of Piperazine in isooctane/AOT/SDS/water microemulsions.

The subscripts interface, water and oil indicate quantities in the surfactant film, water and isooctane respectively, square brackets, as usual, indicate concentrations referred to the total volume of microemulsion, and Z is defined, in analogy with W, as the ratio [isooctane]/([AOT] + [SDS]). Based on these definitions, the model of **Figure 8** implies that the overall pseudo-first-order rate constant is given by Eq. (3):

$$k_o = k_i' \frac{1}{1 + \left(\frac{Z}{K_4}\right)} \quad (3)$$

where k_i' is the pseudo-first-order rate constant in the interface. This constant is expressed in terms of the bimolecular rate constant k_i as:

$$k_i' = k_i \frac{[Amine]_{interface}}{[AOT] + [SDS]} \quad (4)$$

The rate constant at the interface, k_i, is defined in terms of mole per mole concentration in the corresponding microphases. Then Eq. (3) can be rewritten as:

$$k_o = \left(\frac{[Amine]_{TOTAL}}{[AOT] + [SDS]}\right)\left(\frac{k_i K_1 K_4}{(K_1 + W)(K_4 + Z)}\right) \quad (5)$$

The results obtained by fitting Eq. (5) to the experimental data, at each mole ratio ρ, provides the kinetic parameters k_i, K_1 and K_4. **Table 1** shows the values of the rate constant at the interface. The values of the partition constant of the amine, K_1, are independent of the amount of SDS, and are between 9.5 and 10. In addition, the value of the MNTS distribution constant varies between 9 and 12. This result is compatible with those previously obtained in AOT-based microemulsion (K_4 = 11). The average value of K_4 is also in good agreement with the value estimated for isooctane/AOT/water system by analysis of changes in the UV spectrum of MNTS at 265–275 nm [39].

3.2 Reaction with N-methylbenzylamine (NMBA)

The influence of the presence of SDS on the rate of transnitrosation of NMBA by MNTS was studied in a series of experiments in which the total concentration of NMBA was always 0.1 M. The reactions in isooctane/AOT/SDS/water

		PIPERAZINE + MNTS		NMBA + MNTS		MORPHOLINE + MNTS	
ρ	$\overline{V}/M^{-1}$	k_i/s^{-1}	$k_2^i/M^{-1}s^{-1}$	k_i/s^{-1}	$k_2^i/M^{-1}s^{-1}$	k_i/s^{-1}	$k_2^i/M^{-1}s^{-1}$
0.1500	0.336	$(1.57\pm0.07)\cdot10^{-2}$	$5.3\cdot10^{-3}$	$(4.4\pm0.1)\cdot10^{-3}$	$1.5\cdot10^{-3}$	$(8.8\pm0.5)\cdot10^{-4}$	$3.0\cdot10^{-4}$
0.1250	0.341	$(1.64\pm0.08)\cdot10^{-2}$	$5.6\cdot10^{-3}$	$(4.2\pm0.2)\cdot10^{-3}$	$1.4\cdot10^{-3}$	$(9.3\pm0.6)\cdot10^{-4}$	$3.1\cdot10^{-4}$
0.1125	0.344	$(1.55\pm0.05)\cdot10^{-2}$	$5.3\cdot10^{-3}$	$(2.8\pm0.1)\cdot10^{-3}$	$9.5\cdot10^{-4}$	$(8.1\pm0.4)\cdot10^{-4}$	$2.8\cdot10^{-4}$
0.1000	0.347	$(1.56\pm0.06)\cdot10^{-2}$	$5.4\cdot10^{-3}$	-	-	$(8.8\pm0.4)\cdot10^{-4}$	$3.1\cdot10^{-4}$
0.0875	0.350	-	-	$(3.1\pm0.1)\cdot10^{-3}$	$1.1\cdot10^{-3}$	$(8.4\pm0.5)\cdot10^{-4}$	$2.9\cdot10^{-4}$
0.0750	0.353	$(1.57\pm0.06)\cdot10^{-2}$	$5.5\cdot10^{-3}$	-	-	$(8.3\pm0.5)\cdot10^{-4}$	$2.9\cdot10^{-4}$
0.0625	0.356	-	-	$(3.7\pm0.2)\cdot10^{-3}$	$1.3\cdot10^{-3}$	$(8.2\pm0.5)\cdot10^{-4}$	$2.9\cdot10^{-4}$
0.0500	0.359	$(1.50\pm0.05)\cdot10^{-2}$	$5.4\cdot10^{-3}$	$(4.7\pm0.2)\cdot10^{-3}$	$1.6\cdot10^{-3}$	$(8.8\pm0.6)\cdot10^{-4}$	$3.1\cdot10^{-4}$
0.0250	0.364	$(1.50\pm0.06)\cdot10^{-2}$	$5.5\cdot10^{-3}$	$(3.5\pm0.1)\cdot10^{-3}$	$1.3\cdot10^{-3}$	$(8.9\pm0.7)\cdot10^{-4}$	$3.3\cdot10^{-4}$
0.0100	0.368	$(1.47\pm0.08)\cdot10^{-2}$	$5.4\cdot10^{-3}$	$(4.0\pm0.2)\cdot10^{-3}$	$1.5\cdot10^{-3}$	$(8.8\pm0.6)\cdot10^{-4}$	$3.2\cdot10^{-4}$
0.0000	0.370	$(1.45\pm0.08)\cdot10^{-2}$	$5.4\cdot10^{-3}$	$(3.1\pm0.1)\cdot10^{-3}$	$1.2\cdot10^{-3}$	$(6.5\pm0.6)\cdot10^{-4}$	$2.4\cdot10^{-4}$

Table 1.
Kinetic values for the transnitrosation between MNTS and PIP, NMBA, and MOR according to Eqs. (5), (7), and (8).

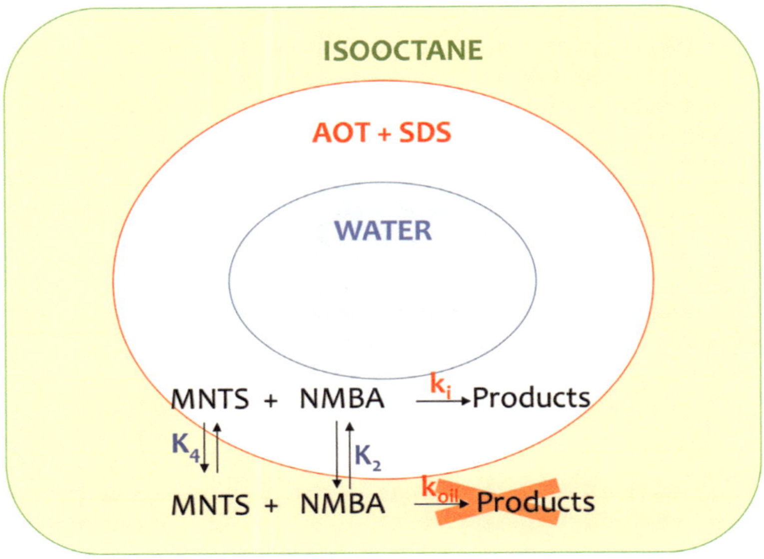

Figure 9.
Pseudophase model for the nitrosation of N-methylbenzylamine in isooctane/AOT/SDS/water microemulsions.

microemulsions were studied by means of series of reactions analogous to those described for piperazine. In this case, no clear trend was observed in the variation of the pseudo-first-order rate constants with W or surfactant concentration.

NMBA is poorly soluble in water. **Figure 9** shows schematically the distribution of the substrates between the continuous medium and the interface of the microemulsion. The pseudophase model for this reaction considers simultaneous reactions in the isooctane and the surfactant film. However, the reaction rate in isooctane (k_{oil}) is several orders of magnitude less than those observed in this system [40–42]. The equilibrium constant K_2 is the partition constant of the amine between the continuous medium and the interface:

$$K_2 = \frac{[Amine]_{interface}}{[Amine]_{oil}} Z \qquad (6)$$

Calculations analogous to those described for piperazine lead to the following expression for the rate constant:

$$k_o = \left(\frac{[Amine]_{TOTAL}}{[AOT] + [SDS]}\right)\left(\frac{k_i K_2 K_4}{(K_2 + Z)(K_4 + Z)}\right) \qquad (7)$$

In this case, K_4 was fixed as 11. Then, the fit of Eq. (7) to the experimental data provides the kinetic parameters k_i and K_2. The values of the partition constant of the amine, K_2, vary between 24 and 28. The value is compatible with the K_2 value obtained previously for the isooctane/AOT/water ternary system (K_2 = 25). **Table 1** shows the values of the rate constant at the interface.

3.3 Reaction with morpholine (MOR)

In experiments on the reaction between morpholine and MNTS, the behavior is similar to that of piprazine: by increasing W the rate constant decreases and by increasing surfactant concentration the rate constant also increases. Morpholine is

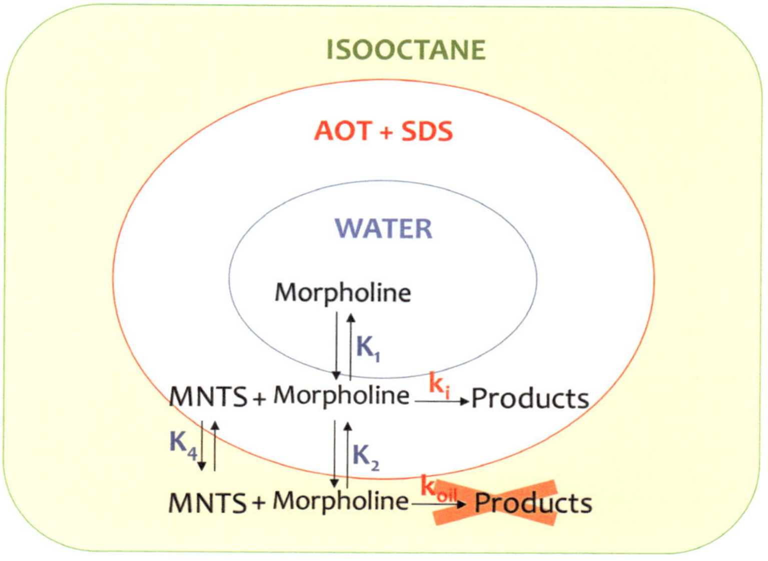

Figure 10.
Pseudophase model for the nitrosation of morpholine in isooctane/AOT/SDS/water microemulsions.

distributed in all three pseudophases in the microemulsion, a more complex situation than any of those considered above. However, the effective reaction region is still constituted by the interface of surfactants, because reaction in isooctane (k_{oil}) is much slower than the observed rates. Taking these considerations into account, the pseudophase model applied is that shown in **Figure 10**.

Calculations analogous to those described for piperazine and N-methylbencilamines lead to the following expression:

$$k_o = \left(\frac{[Amine]_{TOTAL}}{[AOT] + [SDS]}\right)\left(\frac{k_i K_1 K_2 K_4}{(K_1 K_2 + K_2 W + K_1 Z)(K_4 + Z)}\right) \tag{8}$$

The meaning of the constants K_1, K_2 and K_4 was discussed previously. It is considered a known K_4 value equal to 11. The fit of Eq. (8) to the experimental data allows to obtain the other kinetic parameters. The fit, as in the previous cases, is carried out separately for each value of ρ. The distribution constants obtained were K_1 = 45–48 and K_2 = 600. The consistency between the experimental and predicted values (see **Figure 11**) confirms the accuracy of the micellar pseudophase model.

In order to compare the reactivity at the interface of the microemulsion it is necessary to know the molar volume. Furthermore, comparison of reactivity of the amines at the interface with the corresponding reactivity in bulk water needs that k_i (expressed in s^{-1}) must be converted to conventional reaction rates expressed in $M^{-1}\,s^{-1}$. The bimolecular rate constant is:

$$k_2^i = k_i \overline{V} \tag{9}$$

The molar volume of AOT was experimentally determined (given by its density) as 0.37 M^{-1}. For the quaternary system the molar volume can be written as:

$$\overline{V} = \rho \overline{V_{SDS}} + (1 - \rho)\overline{V_{AOT}} \tag{10}$$

Table 1 shows the k_2^i values obtained from the fit of the experimental results to Eqs. (5), (7) and (8) and considering the Eq. (9). The nitroso transfer reactions are 20–50 times slower in isooctane/AOT/SDS/water microemulsions than in bulk water due to the lower polarity of this microregion. The transition state for the transnitrosation reactions of MNTS requires a certain degree of charge separation, and reduction of the polarity will cause a decrease in reaction rate [33]. However, there is no significative difference of the bimolecular rate constants (k_2^i) between isooctane/AOT/water ($\rho = 0$) and isooctane/AOT/SDS/water microemulsions ($\rho = 0.01–0.15$). It is important to remember that in quaternary microemulsions, with $\rho > 0.3$, the kinetics is followed in percolated systems. For piperazine the rate constant is the same ($5.4 \times 10^{-3}\ M^{-1}\ s^{-1}$) in AOT and AOT/SDS microemulsions. For morpholine and NMBA the average value of bimolecular rate constant increases

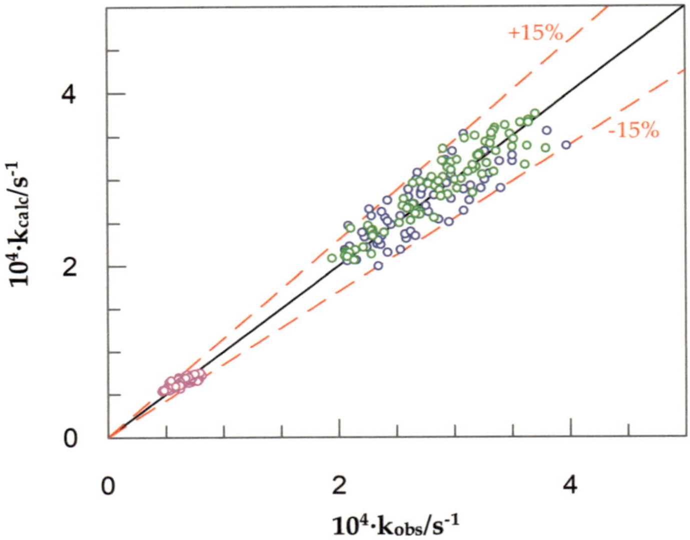

Figure 11.
Plot of experimental vs. calculated rate constants, at 25°C, for the transnitrosation reaction between MNTS and amines in isooctane/AOT/SDS/water microemulsions: piperazine (○), N-methilbenzilamine (○) and morpholine (○).

	Microemulsion	K_1	K_2	K_4	$k_2^i/M^{-1}s^{-1}$	Temperature	Relative k_2^i
Piperazine+MNTS	isooctane/AOT/water	9.6		11	$5.40 \cdot 10^{-3}$	25°C	1.00[a]
	isooctane/AOT/SDS/water	9.6		11	$5.43 \cdot 10^{-3}$	25°C	1.00[a]
	o-xylene/AOT/water	10.6		2.3	$2.94 \cdot 10^{-2}$	50°C	0.66[b]
	isooctane/AOT/water	20.6		7.7	$4.48 \cdot 10^{-2}$	50°C	1.00[b]
NMBA+MNTS	isooctane/AOT/water		25.0	11	$1.20 \cdot 10^{-3}$	25°C	1.00[a]
	isooctane/AOT/SDS/water		25.5	11	$1.33 \cdot 10^{-3}$	25°C	1.11[a]
	o-xylene/AOT/water		20.1	2.3	$4.07 \cdot 10^{-3}$	50°C	0.40[b]
	m-xylene/AOT/water		12.8	2.3	$4.44 \cdot 10^{-3}$	50°C	0.43[b]
	p-xylene/AOT/water		12.5	2.3	$4.81 \cdot 10^{-3}$	50°C	0.47[b]
	isooctane/AOT/water		13.8	7.7	$1.03 \cdot 10^{-2}$	50°C	1.00[b]
Morpholine+MNTS	isooctane/AOT/water	49	600	11	$2.40 \cdot 10^{-4}$	25°C	1.00[a]
	isooctane/AOT/SDS/water	46.4	600	11	$3.03 \cdot 10^{-4}$	25°C	1.26[a]

[a]*Relative* $k_2^i \frac{k_2^i}{(k_2^i)_{isooctane/AOT/water}}$ *at 25°C.*

[b]*Relative* $k_2^i \frac{k_2^i}{(k_2^i)_{isooctane/AOT/water}}$ *at 50°C.*

Table 2.
Kinetic parameter for the transnitrosation between MNTS and PIP, NMBA, and MOR according to Eqs. (5), (7), and (8).

slightly (see **Table 2**). For instance, in the case of morpholine the presence of SDS as cosurfactant increases the k_2^i value from $2.4 \times 10^{-4}\ M^{-1}\ s^{-1}$ to $3.0 \times 10^{-4}\ M^{-1}\ s^{-1}$ (average value). This behavior is the result of two opposite effects: (1) a larger amount of SDS provides a lower molar volume of the surfactant film (see **Table 1**), and (2) by increasing ρ, the polarity at the interface increases. The presence of SDS as cosurfactant increases the small amount of water trapped between surfactant chains. If microemulsions of AOT are compared with those of AOT/SDS, the polarity is likely to increase slightly. For piperazine both effects are compensated and there is no difference in reactivity. The small increase in polarity due the presence of SDS is more important for the amines with solubility in oil: morpholine and NMBA.

4. Nitrosation in xylene/AOT/water microemulsions

The kinetic study of the nitroso group transfer from MNTS to the secondary amines piperazine (PIP) and N-methylbenzylamine (NMBA) in water/AOT/xylene was carried out. For PIP we used only o-xylene/AOT/water microemuslions. For NMBA we used o-xylene, m-xylene and p-xylene as continuous medium. Previously it was shown that these systems are non-percolative. In microemulsions where isooctane provides the continuous medium, W can be as large as 80 at a surfactant concentration of 0.5 M (see **Figure 5**). In contrast, W can barely reach 20 at identical surfactant concentrations in water/AOT/xylene microemulsions at 25°C. Raising the temperature slightly increases the solubility of water in the microemulsion. The kinetic study was conducted at 50°C, above the percolation temperature for isooctane/AOT/water and isooctane/AOT/SDS/water microemulsions.

The influence of the nature of the continuous medium on various properties of a wide range of water-in-oil (w/o) microemulsions was studied in our laboratory [43]. ^{1}H NMR spectroscopy allowed to determine the properties of water in the nanodroplet and the way they are affected by the bulk solvent. Changes in interfacial polarity were examined from the ^{13}C NMR signals of the surfactant molecule AOT. The variation of the carbon chemical shift as a function of the water content (W) was used as a measure of polarity changes at the interface. The reactions of solvolysis of anisoyl chloride were studied in oil/AOT/water microemulsions. The AOT-based microemulsions involved various continuous media, including trichloromethane, tetrachloromethane, alkanes (n-heptane, isooctane, and n-dodecane), cycloalkanes (cyclopentane, cyclohexane, and cycloheptane), and aromatic hydrocarbons (toluene as well as o-xylene, m-xylene and p-xylene). It was found that the solvolysis rate constants of anisoyl chloride depend on the penetration of the oil into the interface.

The kinetic study of PIP reactivity was conducted using o-xylene as the continuous medium in the microemulsions, W values over the range 7.4–15.7 and variable AOT concentrations from 0.1 to 0.7 M. PIP and MNTS concentrations were kept constant at 5×10^{-2} M and 2×10^{-3} M, respectively. The value of pseudo-first-order rate constants (k_o) increase by increasing total surfactant concentration and decreases by increasing the water content (W).

A figure very similar to that used for PIP in the AOT/SDS quaternary system (**Figure 8**) describes the distribution of PIP and MNTS between the three phases of the microemulsion: water, AOT and oil (xylene). Therefore, the pseudo-first-order rate constant will be given by Eq. (5) by substituting [AOT] + [SDS] for [AOT]. Kinetic (k_i) and equilibrium constants (K_1 and K_4) were obtained from fitting experimental data to Eq. (5). The bimolecular rate constant (k_2^i) is estimated from k_i according to Eq. (9) and assuming a value of 0.37 M^{-1} for $\overline{V_{AOT}}$. **Table 2** lists the kinetic parameters for the transnitrosation reaction between PIP and MNTS in o-xylene/AOT/water microemulsion. **Figure 12** compares the experimental rate constants with that calculated. The satisfactory fit obtained for these experiments supports the validity of the model employed.

To compare results, we carried out a kinetic study in isooctane/AOT/water microemulsions at 50°C. Kinetic parameters were determined by following the same procedure as used previously. As can be seen from **Table 2**, the partition equilibrium constant of PIP, K_1, between water and the AOT interface was greater in the o-xylene microemulsions. However, the reaction rate constant at the interphase, k_2^i, was smaller in the interface of the water/AOT/o-xylene microemulsions.

The NMBA reactivity was studied with o-xylene, m-xylene and p-xylene as the continuous medium of the water in oil microemulsion. Nanodroplet size, through the W parameter, varied between 7 and 18 and surfactant concentrations over the range 0.2–0.7 M. NMBA and MNTS concentrations were kept constant at 0.103 M and 2×10^{-3} M, respectively. Pseudo-first order rate constants increased markedly with increasing surfactant concentration at a constant value of W and more slightly with increase in water content at a constant AOT concentration. NMBA and MNTS are virtually insoluble in water and both are distributed between the interface and the oil. The reaction between MNTS and NMBA in xylenes was found to be very slow and, hence, was discarded. Thus, **Figure 9**, is valid for these systems. From the mechanistic proposal shown in **Figure 9** and considering that the surfactant is only AOT, Eq. (7) can be obtained. Kinetics parameters were obtained from fitting experimental data to Eq. (7). The consistency between the experimental and predicted values (see **Figure 12**) confirms the accuracy of the model for xylene/AOT/water microemulsions. **Table 2** compares kinetic and thermodynamic parameters for the microemulsions reviewed in this chapter.

Table 2 shows that the changes in the equilibrium constants (K_1, K_2 and K_4) depend on their distribution between the three pseudophases. The lower values of K_4 in microemulsions with xylenes, at 50°C, may be related with the increased rigidity of the surfactant film due to the penetration of the oil. Similar effect was observed for the distribution constant of the NMBA (K_2). The equilibrium partition

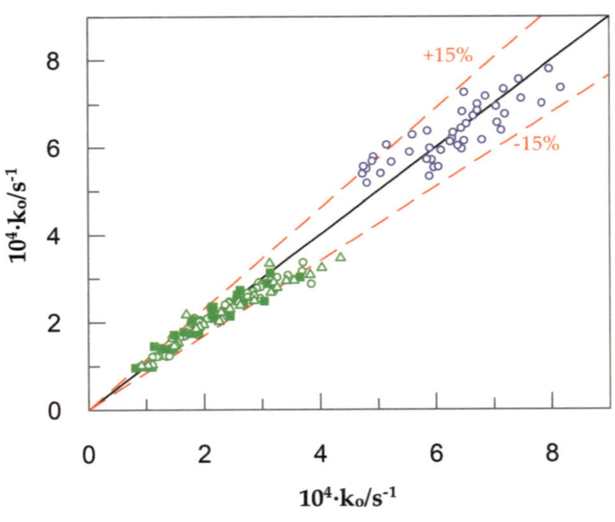

Figure 12.
Plot of experimental vs. calculated rate constants, at 50°C, for the following reactions: PIP + MNTS in o-xylene/AOT/water microemulsions (○), NMBA + MNTS in o-xylene/AOT/water microemulsions (○), NMBA + MNTS in m-xylene/AOT/water microemulsions (■), NMBA + MNTS in p-xylene/AOT/water microemulsions (△).

constant between water and PIP, K_1, exhibits a different behavior; thus, an increase in temperature facilitates the migration of PIP to the pseudo-reaction phase in water/AOT/isooctane microemulsions. This result also seems to confirm the presence of an amount of water at the interphase that increases as the temperature is raised. However, no effect was observed on K_1, K_2 and K_4 for isooctane/AOT/SDS/water microemulsions.

The relative bimolecular rate constant for PIP in o-xylene microemulsions is lower (0.66) than in isooctane microemulsions. This may be a result of a less polar interface by effect of the penetration of the continuous medium. The reaction between MNTS and NMBA was found to be more markedly inhibited in the xylene microemulsions. For NMBA the reaction was 2.1–2.5 times faster in the isooctane microemulsions than in the o-xylene, m-xylene and p-xylene microemulsions. The polarity of the interphase is a function of the amount of oil and water present. Structural evidence suggests that penetration of the continuous medium into the AOT film is easier for the aromatic medium than for isooctane. NMR and kinetic experiments confirm this prediction [43]. Highly polarizable solvents such as benzene, toluene, and tetrachloromethane penetrate the amphiphilic layer presumably up to the water core boundary. The resonance signals for the water H atoms in toluene, m-xylene, and p-xylene microemulsions are suggested an increasing penetration into the oil. The resonance signals for the water H atoms suggest that penetration is slightly deeper with o-xylene than with the other aromatic solvents. The oil penetration increased in the sequence o-xylene > m-xylene > p-xylene [43]. The relative biomolecular rate constants for the reaction between MNTS and NMBA in o-xylene, m-xylene and p-xylene are 0.40, 0.43 and 0.47 respectively.

5. Conclusions

In this chapter we have reviewed the reactions nitrosation of secondary amines by MNTS in AOT-based microemulsions. The reactivity of traditional microemulsions (isooctane/AOT/water) was compared with systems where the interface has been modified by the addition of a cosurfactant or by the substitution of isooctane for xylenes. The presence of SDS as a co-surfactant has important effects on the stability of the microemulsion. An increase in its relative concentration results in a significant decrease in its percolation temperature. The water solubilization capacity also decreases by the presence of SDS. Despite these important effects in the stability of the system, no significant differences are observed in the reactivity. The pseudophase model has been used to quantitatively evaluate the kinetic parameters, but the results are practically the same in isooctane/AOT/SDS/water microemulsions than in isooctane/AOT/water traditional systems.

The substitution of isooctane by xylene gives rise to nonpercolated microemulsions. Xylene/AOT/water microemulsions showed a remarkable effect in the reactivity. The ratio between the bimolecular rate constants allowed to compare the inhibitory effect of isooctane/AOT/water and xylene/AOT/water microemulsions. Thus, the reactions of MNTS with NMBA and PIP were found to be more strongly inhibited in the xylene microemulsions than in the isooctane microemulsions. The increased inhibitory effect of the xylene microemulsions is a result of the decreased polarity at the reaction site: the interphase between the continuous medium and the water.

Finally, we highlight the validity of the kinetic model for percolated and nonpercolated microemulsions. The pseudophase model is independent of structural connotations and internal dynamics of the system and assumes the water, surfactant, and continuous medium to be separate continuous phases.

Acknowledgements

Financial support from Ministerio de Economia y Competitividad of Spain (project CTQ2014-55208-P), Xunta de Galicia (GR 2007/085; IN607C 2016/03 and Centro Singular de Investigación de Galicia Accreditation 2016-2019, ED431G/09), and the European Union (European Regional Development Fund-ERDF) is gratefully acknowledged.

Author details

Pedro Rodríguez-Dafonte
Departamento de Química Física, Facultade de Química and Centro Singular de Investigación en Química Biolóxica e Materiais Moleculares (CIQUS), Universidade de Santiago de Compostela, Santiago, Spain

*Address all correspondence to: pedro.rodriguez@usc.es

References

[1] Luisi PL, Straub BE. Reverse Micelles. New York: Plenum Press; 1984. 354 p. DOI: 10.1002/po.1985.130230312

[2] Tapas KD, Maitra A. Solution behaviour of aerosol OT in non-polar solvents. Advances in Colloid and Interface Science. 1995;**59**:95-193. DOI: 10.1016/0001-8686(95)8005-N

[3] Hoar TP, Schulman JH. Transparent water-in-oil dispersions: The oleopathic hydromicelle. Nature. 1943;**152**:102-103. DOI: 10.1038/152102a0

[4] Schulman JH, Stoeckenius W, Prince LM. Mechanism of formation and structure of micro emulsions by electron microscopy. The Journal of Physical Chemistry. 1959;**63**:1677-1680. DOI: 10.1021/j150580a027

[5] Ekwall P, Mandell L, Fontell K. Some observations on binary and ternary aerosol OT systems. Journal of Colloid and Interface Science. 1970;**33**:215-235. DOI: 10.1016/0021-9797(70)90024-X

[6] Tamamushi B, Watanabe N. The formation of molecular aggregation structures in ternary system: AerosolOT/water/iso-octane. Colloid & Polymer Science. 1980;**258**:174-178. DOI: 10.1007/BF01498277

[7] Gulari E, Bedwell B, Alkhafaji S. Quasi-elastic light-scattering investigation of microemulsions. Journal of Colloid and Interface Science. 1980; **77**:202-212. DOI: 10.1016/0021-9797(80)90432-4

[8] Turco Liveri V, Rossi M, D'Arrigo G, Manno D, Micocci G. Synthesis and characterization of ZnS nanoparticles in water/AOT/n-heptane microemulsions. Applied Physics A. 1999;**69**:369-373. DOI: 10.1007/s003390051016

[9] Calandra P, Goffredi M, Turco LV. Study of the growth of ZnS nanoparticles in water:AOT:n-heptane microemulsions by UV-absorption spectroscopy. Colloids and Surfaces A: Physicochemical and Engineering Aspects. 1999;**160**:9-13. DOI: 10.1016/S0927-7757(99)00256-3

[10] Kim D-W, Oh S-G, Lee J-D. Preparation of ultrafine monodispersed indium-tin oxide particles in AOT-based reverse microemulsions as nanoreactors. Langmuir. 1999;**15**:1599-1603. DOI: 10.1021/LA9815906

[11] Li M, Schnablegger H, Mann S. Coupled synthesis and self-assembly of nanoparticles to give structures with controlled organization. Nature. 1999; **402**:393-395. DOI: 10.1038/46509

[12] Vaucher S, Fielden J, Li M, Dujardin E, Mann S. Molecule-based magnetic nanoparticles: Synthesis of cobalt hexacyanoferrate, cobalt pentacyanonitrosylferrate, and chromium hexacyanochromate coordination polymers in water-in-oil microemulsions. Nano Letters. 2002;**2**: 225-229. DOI: 10.1021/nl0156538

[13] Wong M, Thomas JK, Nowak T. Structure and state of water in reversed micelles. 3. Journal of the American Chemical Society. 1977;**99**:4730-4736. DOI: 10.1021/ja00456a034

[14] Hayes DG, Gomez del Rio JA, Ye R, Urban VS, Pingali SV, O'Neill HM. Effect of protein incorporation on the nanostructure of the bicontinuous microemulsion phase of Winsor-III systems: A small-angle neutron scattering study. Langmuir. 2015;**31**: 1901-1910. DOI: 10.1021/la504606x

[15] García-Río L, Leis JR, Peña ME, Iglesias E. Transfer of the Nitroso Group in Water/AOT/isooctane

microemulsions: Intrinsic and apparent reactivity. The Journal of Physical Chemistry. 1993;**97**:3437-3442. DOI: 10.1021/j100115a057

[16] García-Río L, Leis JR, Mejuto JC. Pseudophase approach to reactivity in microemulsions: Quantitative explanation of the kinetics of the nitrosation of amines by alkyl nitrites in AOT/isooctane/water microemulsions. The Journal of Physical Chemistry. 1996;**100**:10981-10988. DOI: 10.1021/jp953264u

[17] García-Río L, Leis JR. Reactivity in quaternary water in oil microemulsions. 2. Different distribution of the reagents changing from three- to four-component microemulsions. Journal of Physical Chemistry B. 2000;**104**: 6618-6625. DOI: 10.1021/jp000822i

[18] Caponetti E, Lizzio A, Triolo R. Small-angle neutron-scattering study of W/O n-hexadecane, potassium oleate, water and alcohol CnH2n+1OH (n = 5-8) microemulsions: Effect of water concentration. Langmuir. 1990;**6**: 1628-1634. DOI: 10.1021/la00101a002

[19] Caponetti E, Lizzio A, Triolo R, Griffin WL, Johnson JS. Alcohol partition in a water-in-oil microemulsion from small-angle neutron scattering. Langmuir. 1992;**8**: 1554-1562. DOI: 10.1021/la00042a011

[20] Petit C, Bommarius AS, Pileni MP, Hatton TA. Characterization of a four-component cationic reversed micellar system: Dodecyltrimethylammonium chloride/hexanol/n-heptane and 0.1 M potassium chloride solution. The Journal of Physical Chemistry. 1992;**96**: 4653-4658. DOI: 10.1021/j100190a093

[21] Kegel WK, van Aken GA, Bouts MN, Lekkerkerker HNW, Overbeek JTG, de Bruyn PL. Adsorption of sodium dodecyl sulfate and cosurfactant at the planar cyclohexane-brine interface. Validity of the saturation adsorption approximation and effects of the cosurfactant chain length. Langmuir. 1993;**9**:252-256. DOI: 10.1021/la00025a048

[22] García-Río L, Hervella P. Kinetic model for reactivity in quaternary water-in-oil microemulsions. Chemistry —A European Journal. 2006;**12**: 8284-8295. DOI: 10.1002/chem.200600091

[23] García-Río L, Hervés P, Mejuto JC, Pérez-Juste J, Rodríguez-Dafonte P. Pseudophase approach to the transfer of the nitroso group in water/AOT/SDS/isooctane quaternary microemulsions. Langmuir. 2000;**16**:9716-9721. DOI: 10.1021/la000523k

[24] Garcia-Rio L, Hervés P, Mejuto JC, Rodríguez-Dafonte P. Nitrosation reactions in water/AOT/xylene microemulsions. Industrial and Engineering Chemistry Research. 2006; **45**:600-606. DOI: 10.1021/ie050925t

[25] Maitra A, Mathew C, Varshney M. Closed and open structure aggregates in microemulsions and mechanism of percolative conduction. The Journal of Physical Chemistry. 1990;**94**:5290-5292. DOI: 10.1021/j100376a024

[26] Pileni MP. Studies in Physical and Theoretical Chemistry 65: Structure and Reactivity in Reverse Micelles. Amsterdam: Elsevier; 1989. 379 p

[27] García-Río L, Leis JR, Mejuto JC, Peña E. Effects of additives on the internal dynamics and properties of water/AOT/isooctane microemulsions. Langmuir. 1994;**10**:1676-1683. DOI: 10.1021/la00018a013

[28] Alvarez E, García-Río L, Mejuto JC, Navaza JM, Pérez-Juste J, Rodríguez-Dafonte P. Effect of temperature on the electrical conductivity of sodium bis(2-ethylhexyl)sulfosuccinate

+ 2,2,4-trimethylpentane + water microemulsions. Influence of alkylamines. Journal of Chemical & Engineering Data. 1999;**44**:1286-1290. DOI: 10.1021/je990108y

[29] Atik SS, Thomas JK. Transport of photoproduced ions in water in oil microemulsions: Movement of ions from one water pool to another. Journal of American Chemical Society. 1981; **103**:3543-3550. DOI: 10.1021/ja00402a048

[30] García-Río L, Hervés P, Leis JR, Mejuto JC. Influence of crown ethers and macrocyclic kryptands upon the percolation phenomena in AOT/isooctane/H_2O microemulsions. Langmuir. 1997;**13**:6083-6088. DOI: 10.1021/la970297n

[31] Ray S, Paul S, Moulik SP. Physicochemical studies on microemulsions: V. Additive effects on the performance of scaling equations and activation energy for percolation of conductance of water/AOT/heptane microemulsion. Journal of Colloid and Interface Science. 1996;**183**:6-12. DOI: 10.1006/jcis.1996.0512

[32] Castro A, Leis JR, Peña ME. Decomposition of N-methyl-N-nitrosotoluene-p-sulphonamide in basic media: Hydrolysis and transnitrosation reactions. Journal of the Chemical Society, Perkin Transactions 2. 1989;**0**:1861-1866. DOI: 10.1039/P29890001861

[33] García-Río L, Iglesias E, Leis JR, Peña ME, Rios A. Reactivity of nucleophilic nitrogen compounds towards the nitroso group. Journal of the Chemical Society, Perkin Transactions 2. 1993;**0**:29-37. DOI: 10.1039/P29930000029

[34] Bravo C, Hervés P, Leis JR, Peña ME. Micellar effects in the acid denitrosation of N-nitroso-N-methyl-p-toluenesulfonamide. The Journal of Physical Chemistry. 1990;**94**:8816-8820. DOI: 10.1021/j100388a014

[35] Bravo C, Leis JR, Peña ME. Effect of alcohols on catalysis by dodecyl sulfate micelles. The Journal of Physical Chemistry. 1992;**96**:1957-1961. DOI: 10.1021/j100183a077

[36] García-Río L, Leis JR, Mejuto JC, Pérez-Juste J. Investigation of micellar media containing β-cyclodextrins by means of reaction kinetics: Basic hydrolysis of N-methyl-N-nitroso-p-toluenesulfonamide. The Journal of Physical Chemistry. B. 1997;**101**:7383-7389. DOI: 10.1021/jp970862v

[37] Hervés P, Leis JR, Mejuto JC, Pérez-Juste J. Kinetic studies on the acid and alkaline hydrolysis of N-methyl-N-nitroso-p-toluenesulfonamide in dioctadecyldimethylammonium chloride vesicles. Langmuir. 1997;**13**:6633-6637. DOI: 10.1021/la9705975

[38] García-Río L, Hervés P, Leis JR, Mejuto JC, Perez-Juste J. Hydrolysis of N-methyl-N-nitroso-p-toluenesulfonamide in micellar media. Journal of Physical Organic Chemistry. 1998;**11**:584-588. DOI: 10.1002/(SICI)1099-1395(199808/09)11:8/9<584::AID-POC59>3.0.CO;2-F

[39] García-Río L, Hervés P, Mejuto JC, Pérez-Juste J, Rodríguez-Dafonte P. Pseudophase approach to reactivity in microemulsions: Quantitative explanation of the kinetics of the nitroso group transfer reactions between N-methyl-N-nitroso-p-toluenesulfonamide and secondary alkylamines in water/AOT/isooctane microemulsions. Industrial and Engineering Chemistry Research. 2003;**42**:5450-5456. DOI: 10.1021/ie0208523

[40] García-Río L, Leis JR, Iglesias E. Nitrosation of amines in nonaqueous solvents. 1. Evidence of a stepwise mechanism. The Journal of Organic

Chemistry. 1997;**62**:4701-4711. DOI: 10.1021/jo970188d

[41] García-Río L, Leis JR, Iglesias E. Nitrosation of amines in nonaqueous solvents. 2. Solvent-induced mechanistic changes. The Journal of Organic Chemistry. 1997;**62**:4712-4720. DOI: 10.1021/jo9701896

[42] Boni JC, García-Río L, Leis JR, Moreira JA. Nitrosation of amines in nonaqueous solvents. 3. Direct observation of the intermediate in cyclohexane. Journal of Organic Chemistry. 1999;**64**:8887-8892. DOI: 10.1021/jo991136m

[43] García-Río L, Godoy A, Rodríguez-Dafonte P. Influence of the oil on the properties of microemulsions as reaction media. European Journal of Organic Chemistry. 2006;**2006**:3364-3371. DOI: 10.1002/ejoc.200600086

Section 2

Nanotemplates and Nanomaterials Synthesis

Chapter 3

Microemulsions as Nanotemplates: A Soft and Versatile Approach

Rohini Kanwar, Jyoti Rathee, Madhuri Tanaji Patil and Surinder Kumar Mehta

Abstract

Template efficacy of microemulsions in generating nanoparticles has garnered considerable attention in the world of colloidal science. A microemulsion is an optically isotropic and thermodynamically stable colloidal dispersion, which possess spherical droplets (either of W/O or O/W) of the size <50 nm. In microemulsions, the spontaneous formation of domains of nanometric dimensions significantly facilitates their exploitation as potential nanoreactors for the production of stable nanoparticles (due to their cost-effectiveness and ease of preparation). The present chapter provides an overview of microemulsions as efficient nanotemplates, with a detailed account of plausible nanomaterials, i.e., metallic nanoparticles, quantum dots, polymeric nanoparticles, mesoporous silica nanoparticles, solid lipid nanoparticles, nanostructured lipid carriers, etc. Based on the high surface area, good crystallinity, controllable particle size, outstanding catalytic, and magnetic properties, the exploitation of nanoparticles as efficient catalysts and drug delivery modules has also been highlighted.

Keywords: microemulsions, nanotemplates, metallic nanoparticles, quantum dots, polymeric nanoparticles, mesoporous silica nanoparticles, solid lipid nanoparticles, nanostructured lipid carriers

1. Introduction

Microemulsions are thermodynamically stable, optically isotropic and spontaneously formed colloidal dispersed system of two immiscible liquids (such as oil and water), which are stabilized by the interfacial film of a surfactant (and co-surfactant) [1]. The formation of microemulsion was first described by Hoar and Shulman in 1943, after observing a spontaneous and well-defined transformation of an opaque emulsion to a transparent solution upon addition of medium chain alcohol (co-surfactant) [2]. However, it was the year 1959, when Shulman et al. [3] coined the term called "microemulsion." A microemulsion is a macroscopically homogeneous system and possesses spherical droplets of the size <50 nm that do not require the higher input of energy and shear reaction conditions, in contrast to conventional emulsions, which are cloudy, kinetically stable and thermodynamically unstable systems [4]. Microemulsions can be considered akin to micellar solutions that solubilize the oil domain into the nonpolar surfactant tail region to give stable microstructure.

Basically, a microemulsion comprises of three components, namely a polar phase (water), a nonpolar phase (oil) and an emulsifier. On a microscopic level, the emulsifier molecules form an interfacial film, which separates the polar and the nonpolar domains. The formed interfacial layer leads to different microstructures ranging from oil-swollen direct micelles dispersed in water (O/W microemulsion) over a bicontinuous "sponge" to water-swollen inverse micelles dispersed in oil (W/O microemulsion) phase. Depending on the hydrophilic-lipophilic balance (HLB) value of the surfactant and the oil/water ratio, the formed microstructure can exist in oil-in-water (O/W), water-in-oil (W/O), hexagonal, reverse hexagonal, or a mixture of (O/W and W/O) called bicontinuous/lamellar phase. In general, the surfactants having HLB of 3–6 promote the formation of W/O microemulsions whereas with HLB of 8–18 facilitate the formation of O/W microemulsions [5]. The W/O microemulsion is generally termed as "reverse micelles" in which water swollen micelles are dispersed into the oil phase, and polar head groups of the surfactant are attracted by the aqueous phase, whereas the hydrocarbon chain is attracted by the oil phase. The reverse micelles obtained are of spherical shape, monodispersed that can easily control the size of the aqueous core up to 5–10 nm. On the other hand, in O/W microemulsion, the size of droplets can be tailored up to 1–100 nm by varying the concentration of the dispersed phase and the surfactants.

Figure 1 shows the ternary phase diagram, the three edges of which represents the components of a microemulsion, namely, oil, water, surfactant (and a co-surfactant, referred as a pseudocomponent is added).

In literature, different theories are documented illustrating the spontaneous formation of microemulsion such as interfacial, solubilization and thermodynamic theories, etc., out of which few main theories are described below. In 1955, the first theory, i.e., mixed film theory was proposed, which considered the interfacial film as a duplex film, i.e., interface is the third phase and have two-dimensional region bounded by oil on one side and water on the other [6]. It was postulated that the spontaneous formation of the microemulsion is attributed to the interactions at the interphase, where the interfacial tension between oil and water phase is brought down to zero. However, only on the basis of molecular interactions across the duplex film, the formation of microemulsion could be ensured rather than other liquid crystalline phases, in which one bulk phase gets enclosed in the other (in the form of spheres). Based on the mixed film theory, Robbins [7] devised a theory of phase behavior of microemulsions, which stated that interactions in a mixed film direct the direction and extent of curvature, by which the type and size of droplets

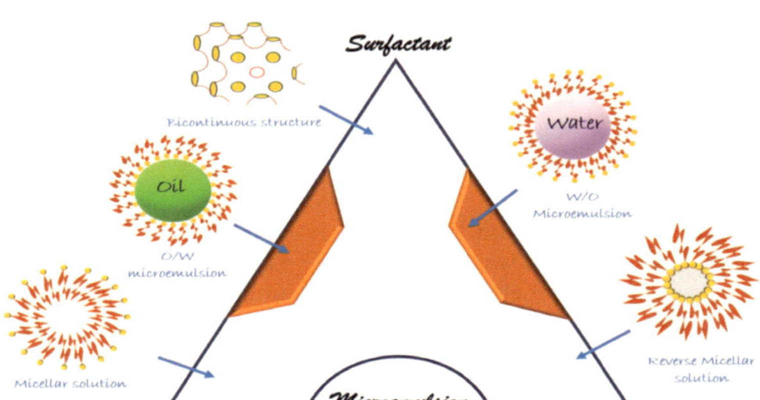

Figure 1.
Ternary phase diagram representing three components of the microemulsion.

of the microemulsions can be estimated. Later, the solubilization theory was proposed, which regarded microemulsions as the swollen micellar systems. In 1969, Adamson [8] gave a model in which the W/O emulsion was said to be formed due to the balancing of Laplace and osmotic pressure. The thermodynamic theory showcased that the free energy of formation of microemulsions consists of interfacial energy and energy of the clustering droplets. It is the reduction in the interfacial free energy that facilitates the spontaneous formation of the thermodynamically stable microemulsion. Also, with the thermodynamic approach, the information related to the stability and size of droplets in microemulsion can be deduced. Another important theory was given by Schulman and his co-workers [3], who reported that the negative interfacial tension is a transient phenomenon for the spontaneous uptake of water or oil in microemulsion and it is the interfacial charge, which controls the phase continuity.

Based on the positive attributes of microemulsions such as spontaneity, thermodynamic stability, and solubilization potential (illustrated by the theories), the developed assembly acts as a potential template for the fabrication of diverse nanoparticles.

2. Microemulsions as efficient nanotemplates

Microemulsion, a soft and versatile approach, has the distinct ability to modulate the particle properties such as the morphology, geometry, surface properties (activity and selectivity), surface area, stability, and homogeneity of the formed nanoparticles (NPs). It is a unique method to regulate the kinetics of the NPs formation and growth (controlled by thermodynamic stabilization with surfactant molecule) by altering the physiochemical properties of the microemulsion. Microemulsions are dynamic systems, in which formed droplets frequently collide, coalesce with each other due to the continuous Brownian motion. They tend to merge among themselves to form transient dimers, which break apart the surfactant layer, and thereby induce the micellar exchange within the interior of the droplets. The droplet content exchange phenomena occur in the order of millisecond to microsecond time scale. From this, it can be inferred that the microemulsion can act as an efficient "nanoreactors" or excellent reaction site, which can facilitate the synthesis of diverse NPs (owing to its dynamicity). These surfactant-covered water pools/oil pools offer a unique microenvironment for the formation of NPs (where the surfactant layer prevents the NPs from aggregation by its steric stabilization property) [9].

2.1 Basics of NPs formation

Microemulsions can act as efficient templates to synthesize NPs via. Two routes: (A) one microemulsion method and (B) two microemulsion method [10]. **Figure 2** shows the mechanism involved behind the formation of NPs from the microemulsion methods. In one microemulsion method, a triggering agent is a prerequisite to initiate the nucleation reaction, which can be either present within the single microemulsion (containing the precursor) or added externally into the microemulsion (as a second reactant) (**Figure 2A**). For the synthesis of NPs, the triggering reactant has to diffuse through the interfacial wall of the microemulsion, i.e., why one microemulsion method is a diffusion-controlled process.

However, in the two microemulsion methods, the two microemulsions (consisting of separate reactants) are mixed together in suitable ratios, in which the Brownian motion of the micelles brings them in contact with each other. Because of

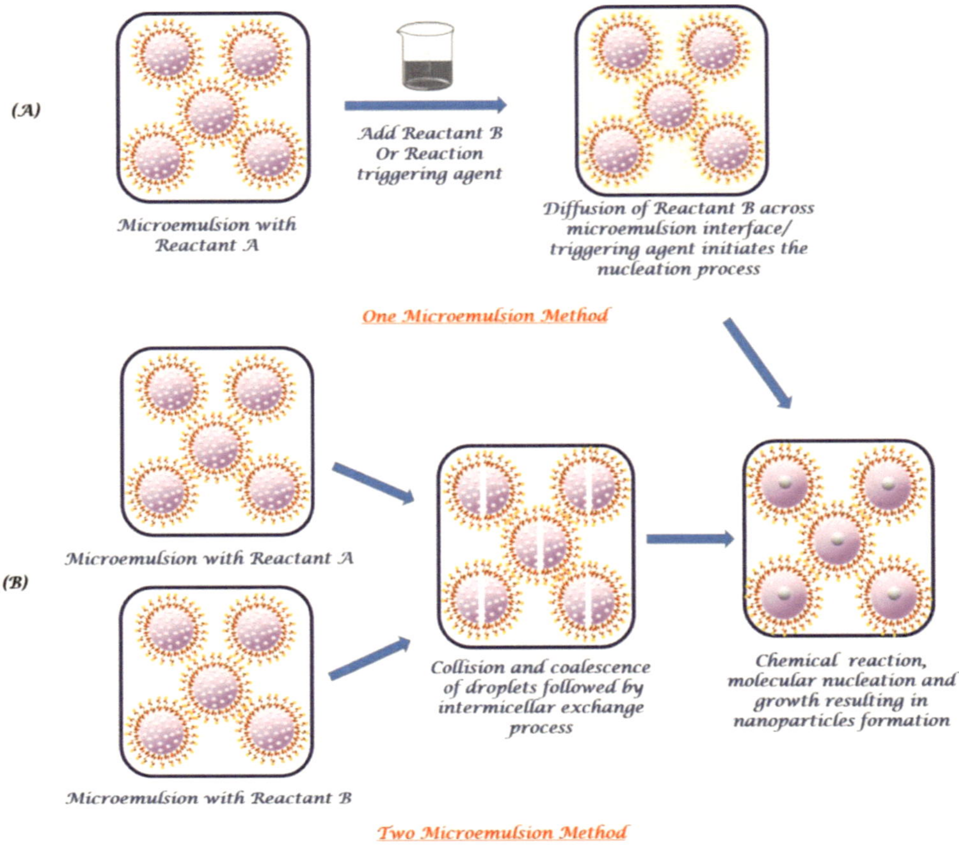

Figure 2.
Mechanism involved behind nanoparticles preparation from microemulsion method: (A) mixing of two microemulsions and (B) direct addition of reducing agent to the microemulsion.

which intermicellar collisions and sufficiently energetic collisions happen, thereby, leading to the mixing of the micellar components (**Figure 2B**). The chemical reaction happens, when both the reactants come in the same vicinity. And when the critical number of molecules is attained inside the micellar units, the nucleation process progresses followed by growth and coagulation of primary particles, resulting in the NP formation.

2.2 Different types of NPs (developed from the microemulsion method)

Microemulsions are potent chemical nanoreactors with distinct interfacial properties providing which provides an intimate contact of hydrophilic and hydrophobic domains at the nanoscale level. By utilizing microemulsions, various nanomaterials have been synthesized such as metallic NPs, quantum dots, polymeric NPs, mesoporous silica NPs, solid lipid NPs, nanostructured lipid carriers, etc. (**Figure 3**).

2.2.1 Metallic nanoparticles

Metallic NPs such as silver (Ag), gold (Au), platinum (Pt), palladium (Pd), copper (Cu), nickel (Ni), molybdenum (Mo), ruthenium (Ru), selenium (Se), iron (Fe), cerium (Ce) and their oxides, sulfides, fluorides, chromates, phosphates, etc.

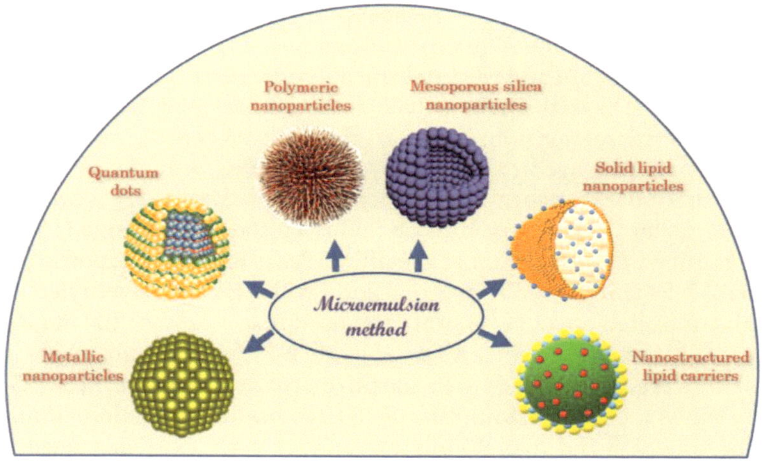

Figure 3.
Structure of nanoparticles (NPs) prepared using microemulsion method.

have drawn significant attention because of their emerging applications in the field of catalysis, clinical diagnostics, and therapy. For the synthesis of metallic NPs using microemulsion, two schemes have been identified [11]: (a) O/W microemulsion in which ionic salt (metallic precursor) as the precursor is dissolved in the continuous aqueous phase and (b) W/O approach in which organometallic salt (metallic precursor) is dissolved in the oil phase of microemulsion.

In 1982, for the first time, Boutonnet et al. [12] recognized microemulsions as the convenient template for the synthesis of metal NPs using W/O microemulsion. They synthesized the monodispersed Pt, Pd, Rh, and Ir NPs within the size range of 2–5 nm by reducing the corresponding salt of these metals in water pool of O/W emulsion (by hydrogen gas). They showed that in order to achieve the stability and minimize the water content of the microemulsion, a high concentration of reducing agent is needed. Following the pioneering work of Boutonnet, the exploitation of microemulsion for the synthesis of NPs gained considerable momentum in the scientific arena [11, 13, 14]. Pileni and Lisiecki [14] prepared Cu NPs using sodium dioctyl sulfosuccinate (AOT)-based W/O microemulsion, where they changed the shape of NPs by varying the ratio of water/surfactant in the microemulsion system. Tianimoghadam and Salabat [15] fabricated monodispersed thiol-functionalized Au NPs (having diameter 3–4 nm) by using W/O microemulsion (toluene/tetraoctylammonium bromide/water) where $HAuCl_4$ was employed as a precursor and $NaBH_4$ as a reducing agent. Perez-Coronado and co-workers [16] synthesized Pd NPs using W/O microemulsion (AOT/iso-octane system/water) and used them as a catalyst for reduction of bromate with H_2 in water. An inverse microemulsion method was utilized to fabricate Ni, Mo, Fe NPs (of size 2.0–2.5 nm and agglomerates <50 nm having spherical morphology) at room temperature and exploited to heavy crude oil *in situ* hydroprocessing and to enhance their physiochemical properties [17]. Yamagishi et al. [18] established that it is the interaction between the metal salt and micellar components, which directs the capacity of the microemulsion to solubilize the metal salts in it. Furthermore, Destrée and Nagy [13] as well as Lopez-Quintela [19] proposed the mechanism of NPs formation using W/O microemulsion.

Later, in 2005, Ge et al. [20] described the potential of O/W microemulsion for the preparation of different inorganic NPs such as metal NPs (Cu, Ag), fluorides (CaF_2, YF_3, PrF_3, NdF_3), semiconductors (ZnS, Ag_2S, CdS, PbS, CdSe, PbSe), chromates ($BaCrO_4$ and $PbCrO_4$), and phosphates ($CePO_4$ and $HoPO_4$) in the size range of 2–13 nm. They proposed that the metal cation gets adsorbed at the oil-water interface of the preformed O/W microemulsion because of the Coulombic attraction between the metal ion and the linoleate anion from the surfactant. Due to the strong solvation of ions in the polar solvents, they favor being positioned at the interface of O/W microemulsion; however, this balance gets destroyed upon addition of the precipitating agent, and thereby, leads to the formation of smaller sized particles. Furthermore, Sanchez-Dominguez and co-workers [21] synthesized Pt, Pd, or Rh NPs using O/W microemulsion and from the color change (from gray to brown) determined the formation of metallic NPs. Li et al. [22] reported the synthesis of Co-B NPs using O/W microemulsion (cyclohexane/polyethylene glycol/water) with controllable size (6–20 nm) and used it as a catalyst for the hydrogenation process. Ce oxide and mixed Cu/Ce oxide NPs (size 2–3 nm) with low polydispersity were synthesized by Pemartin-Biernath and co-workers [23]. Furthermore, a novel method to synthesize Fe NPs *in situ* in O/W microemulsion (Brime/n-hexane/SDS and span 80 mixture/isopropyl alcohol) was developed by Hu et al. [24] to increase their performance in oil recovery. Rivera-Rangel and co-workers [25] fabricated the Ag NPs by using the O/W microemulsion (Brij 96/1, 2-hexanediol/castor oil, extract of geranium leaf as a reducing agent) method as a template. They found that in the microemulsion method, the obtained Ag NPs were of a controllable size and the method was more eco-friendly as compared to the nanoemulsion method.

Among these two approaches, O/W is found more convenient and an environment-friendly approach (having water as the major (continuous) phase) as compared to W/O, which includes the use of oil in a larger amount resulting in a small yield of NPs per microemulsion volume and hindering the applications at industrial scale.

2.2.2 Quantum dots

In the last few decades, scientists have shown enormous interest in the field of nanostructured materials called "quantum dots." Quantum dots (QDs) are inorganic nanocrystals that act as fluorophores and finds plenty of applications in electronics, biosensing, *in vivo* imaging, chemical, and biomedical researches. QDs are basically zero-dimensional entities consisting of a semiconductor core (<30 nm size), coated by a shell and an organic cap or an inorganic layer (a material having larger band gap), which enhances its solubility in aqueous buffer solutions [26]. So far, a wide range of quantum dots has been synthesized such as CdX (X = S, Se, Te), carbon dots, Si QDs, graphene QDs, Ag_2Se, Ag_2S, ZnS, InP, etc., because of their good biocompatibility and excellent optical properties [27, 28]. Optoelectric properties of QDs were found directly dependent upon the size and shape of QDs, as large quantum dots (5–6 nm size) emit longer wavelength with an emission color of orange/red, whereas smaller quantum dots (2–3 nm size) emit a small wavelength of blue/green color [29].

Among the various methods, the reverse microemulsion technique (the bottom-up approach) is considered as the most convenient and popular method approach, as one can easily tune the size of QDs (by altering the surfactant/water molar ratio) and NPs with narrow size distribution can be prepared. To synthesize the QDs, the nanometric-sized water droplets are dispersed in n-alkane solutions using surfactants like AOT, cetyltrimethylammonium bromide (CTAB), sodium dodecyl sulfate (SDS), or triton-X. Shakur [30] fabricated ZnS QDs by reverse microemulsion method (using pyrrolidone as the surfactant) having the cubic structure of size 2.1 nm, the lattice parameter of 5.4 Å, and specific surface area of 1.81 cm^2/g. Yang and coworkers [31] synthesized ZnSe and Fe-doped ZnSe QDs using W/O microemulsion hydrothermal technique of spherical shape having zinc blende structure and monodispersed nature. CdS QDs of 3.8 nm size and cubic phase nanocrystals were developed by using reverse microemulsion (cetylpyridinium chloride/1-pentanol/water/heptane, and $CdCl_2$ and Na_2S) at 303 K [32]. The microemulsion-mediated hydrothermal method was utilized by Chen and co-workers [33] in which CdS QDs with controllable size and crystallinity were fabricated by a chemical reaction of cadmium acetate dehydrate and thioacetamide CdS. Tarkas et al. [34] synthesized monomorphic SnS QDs using the surfactant-free microemulsion (chlorobenzene/methanol/ethylene glycol). They showed that the microemulsion concentration (primary factor) and microemulsion temperature (secondary factor) are the important parameters that can influence the diameter of QDs.

2.2.3 Polymeric nanoparticles

Polymeric nanoparticles (PNPs) have attracted substantial attention due to their distinct optical, electrical, optoelectrical attributes and interesting applications in biomedical sciences, catalysis, sensing, etc. [35]. PNPs are defined as solid NPs or particulate dispersions, which are prepared from biocompatible and biodegradable polymers in the size range of 10–1000 nm [36]. Generally, the preparation of PNPs include two main steps: (1) preparation of emulsified system (emulsions or microemulsions or nanoemulsions) and (2) secondly, the formation of NPs either by the precipitation or polymerization of the monomers or the gelation of polymer. For the synthesis of PNPs, the most commonly exploited natural polymers are chitosan, gelatin, sodium alginate, and albumin and synthetic polymers are polyactides, polyaniline (PAni), polyglycolides, poly(lactic co-glycolides), polyorthoesters, polycaprolactone, polyglutamic acid, polymalic acid, poly(vinyl alcohol), poly(methyl methacrylate) (PMMA), and many more [37].

Microemulsion polymerization is regarded as one of the convenient and effective approach for the preparation of PNPs (producing colloidal polymer particles of high molar mass) in order to modulate the structural properties and obtain the smaller and narrow size distribution of PNPs. Briefly, an initiator (water soluble) is added to the aqueous phase (of the thermodynamically stable microemulsion) comprising swollen micelles, which initiates the polymerization process. Due to the utilization of a high amount of surfactant, the formed PNPs get completely covered with the surfactant, which enhances the stability. Initially, in some droplets, the polymer chains are formed (as simultaneous initiation in all the microdroplets is not possible). Furthermore, it is the osmotic and elastic influence of the polymeric chains that destabilizes the susceptible microemulsions resulting in the increased particle size, empty micelles formation, and secondary nucleation. The kinetics of microemulsion polymerization and properties of PNPs are entirely dependent on the type of initiator and its concentration, monomer, surfactant, and reaction temperature [36].

In the early 1980s, Stoffer and Bone [38] reported the first microemulsion polymerization of methyl methacrylate using W/O microemulsion yielding stable polymer latexes of the size 10–100 nm. Furthermore, Ming and co-workers [39] used low surfactant/monomer ratio of 0.09 (w/w%) and developed 6–24 (w/w%) PMMA latex of the size 33–46 nm. Thereafter, Gan et al. [40] and Selvan et al. [41], for the first time reported the preparation of PAni NPs by employing (sodium bis(2-ethylhexyl) sulfosuccinate (AOT)-water-cyclohexane) microemulsion. However, the polymerization rate was slow and the morphology of PNPs was not uniform. So, as to overcome this problem long-term stable polymer/nanosilica composite latex was prepared under ultrasonic irradiation [42, 43]. A new method: ultrasonic assisted inverse microemulsion (CTAB/n-hexanol/concentered HCl) polymerization method was developed to prepare conducting PAni NPs with uniform shape and size ranging between 10 and 100 nm [35]. Polyacrylamide NPs (with size <100 nm) were prepared using the reverse micelle (Aerosol/n-hexane/acrylamide/N,N′ methylene bisacrylamide) by Munshi and co-workers [44]. Chitosan magnetic NPs were prepared *in situ* within the size range of 10–80 nm in the microreactor of tiny water pools of W/O microemulsion (Triton X-100/cyclohexane/n-hexanol/hydrochloric acid) [45]. Ge and co-workers [46] prepared the biodegradable NPs by precipitating the W/O microemulsion (RNA aqueous solution/dichloromethane (poly(L-lactic acid)-poly(ethylene glycol)/n-octyl β-D-glucopyranoside/n-butanol) in supercritical CO_2. Ahmad et al. [47] prepared chitosan coated poly-lactide-*co*-glycolide NPs using O/W microemulsion system and reported better results (entrapment efficiency, *in vitro* release and *in vitro* cytotoxicity) than conventional drug solution and unloaded NPs.

2.2.4 Mesoporous silica nanoparticles

Advancement in the field of mesoporous silica nanoparticles (MSN) has increased dramatically owing to its various advantages such as large surface area and pore volume (which makes its potential candidate for drug adsorption and loading within the pore channels), excellent mesoporous structure and adjustable pore size, and easily modifiable surface structure. Initially, MSN were synthesized and reported by the research groups of Cia, Mann, and Ostafn; however, after the work of Victor Lin, the term MSN became more familiar [48]. There are various methods reported for the synthesis of MSN such as sol-gel process, reverse microemulsion, flame synthesis, and many more. But among those methods, reverse microemulsion method is the most widely exploited method. Briefly, the surfactant molecules are dissolved in the organic solvents to form the spherical micelles, and in the presence of water, the polar head group of the surfactant organizes itself to form microcavities containing water called "reverse micelles." In microemulsion, the nucleation and growth of particles are restricted within the water core of the inverse micelles that leads to the synthesis of MSN with specific size and morphology. Furthermore, there are two prerequisite conditions that need to be taken into account during the synthesis of MSN, i.e., firstly, the well-controlled nucleation growth rate of MSN and secondly, the nonsticky nature of MSN [49].

The common type of mesoporous silica includes 2D hexagonal mobile crystalline material (MCM)-41 and a 3D cubic Santa Barbara amorphous (SBA)-15 having a pore size between 2 and 10 nm. Tetraethyl orthosilicate (TEOS) is exploited as the common source of silicon alkoxide (by adding into the reverse microemulsion) for the synthesis of spherical silica NPs in the desired size range. Finnie and co-workers [50] synthesized the MSN by the reaction of TEOS inside the water droplets of W/O microemulsion (nonylphenylether/cyclohexane/water) under both acidic and basic conditions, and controlled the NPs formation, optimized the yield and the release rate of the encapsulated drug. Furthermore, Kao and Mou [51] reported the synthesis of MSN using O/W microemulsion (CTAB/decane/ethanol) having a tunable pore size and showed high adsorption capacity of MSN for the lysozyme. Shang et al. [52] fabricated the MSN of reverse bumpy ball structure (RBBS) by using O/W microemulsion (CTAB/bromide/polydecane/cyclohexane) and TEOS. And furthermore, they established that the RBBS containing Pt-γ-Fe_2O_3 dimers shows an excellent catalytic performance for the reduction of *p*-nitrophenol (by H_2).

2.2.5 Solid lipid nanoparticles

Solid lipid nanoparticles (SLNs) are termed as first generation of lipid NPs, which are developed to overcome the disadvantages associated with the emulsions (such as drug degradation, high drug payload, etc.) by replacing the liquid state of the lipid with the solid state (both at room temperature and body temperature) of the lipid. In the year 1990, the research group of Gasco (Turin, Italy), Müller (Berlin, Germany) and Westesen (Braunschweig, Germany) actively engaged in developing the SLNs corresponding to the size range of 10–1000 nm. In literature, different methods are documented for the fabrication of SLNs; however, the microemulsification method is regarded as the most promising route for the synthesis of SLNs because of its simplicity, cost effectiveness, enhanced drug stability, high drug payload, and controlled drug release. The most commonly exploited solid lipids are triacylglycerols, triglycerides, acylglycerols, fatty acids, waxes, and others. Basically, the microemulsion method involves two steps: (i) the melted lipid matrix is dispersed into an aqueous surfactant solution under constant stirring and a hot microemulsion is developed and (ii) then it is quenched into cold water (2–3°C) to solidify the lipid droplets [53].

There exists an extensive literature for SLNs prepared using microemulsion method. For the first time, Morel and co-workers [54, 55] employed SLNs to encapsulate peptide drugs ([D-Trp-6] LHRH and thymopentin) using W/O/W microemulsion-based technique. Then, Ugazio and co-workers [56] incorporated the hydrophobic peptide using W/O microemulsion technique (with a matrix comprising of stearic acid, phosphatidylcholine, taurocholate; encapsulating up to 13% of cyclosporine A) and prepared cyclosporine A loaded SLNs. Furthermore, SLNs consisting of excipients: stearic acid, emulsifying wax, octadecyl alcohol, and cetyl palmitate were fabricated and employed for delivery of anticancer drugs [57]. Kuo and Chung [58] formulated SLNs having spheroidal morphology with shallow surface pits using the complex core of Compritol 888 ATO, tripalmitin, and cacao butter, and finally, targeted for antiretroviral drugs like stavudine, delavirdine, and saquinavir. Mehrad and co-workers [59] prepared SLNs having size <220 nm with spherical morphology using microemulsification method. They employed palmitic acid as the solid lipid and stabilized it with whey protein isolate (surfactant) for enhancing the physicochemical stability of β-carotene by encapsulating into SLNs. However, recently, Kanwar et al. [60] exploited the potential of SLNs as an efficient template for synthesizing the CuO-embedded meso-macroporous silica framework. They employed the ascribed catalyst for the reduction of *p*-nitrophenol to *p*-aminophenol and hexacyanoferrate (III) ions to hexacyanoferrate (II) ions, which showed promising results in the field of nanocatalysis than other available catalysts.

2.2.6 Nanostructured lipid carriers

Nanostructured lipid carriers (NLCs) are considered as the second generation of lipid NPs, which are developed to overcome the problems associated with SLNs (such as increasing the loading capacity and preventing the drug expulsion). In the year 1999, Müller et al. introduced NLCs comprising a mixture of solid lipid and liquid lipid in order to create more imperfection in the matrix of NLCs. Following the similar procedure of SLNs, NLCs were developed mainly by the microemulsion method. Doktorovova et al. [61] fabricated NLCs using microemulsion method (with excipients: precirol ATO 5, labrasol, Tween 80, and soybean lecithin), which exhibited excellent stability and entrapment efficiency for fluticasone propionate. Kuo and Chung [62] synthesized NLCs of 160 nm size, with uniform size distribution using excipients: stearic acid, Compritol 888, oleic acid, and Tween 80 and reported the efficient delivery of nevirapine for viral therapy. Khurana et al. [63] fabricated the meloxicam-loaded NLCs using microemulsion template strategy and showed the sustained release of meloxicam from SLNs. Shao and coworkers [64] synthesized the transferrin-decorated NLCs for the delivery of the paclitaxel by microemulsion technique (glyceryl monostearate/oleic acid/soy lecithin). Kanwar et al. [65] for the first time employed the cationic lipid (didodecylammonium bromide) as the core lipid and formulated its cationic NLCs (of the size 160 nm and >30 mV zeta potential) using microemulsification method. Chanburee and Tiyaboonchai [66] fabricated curcumin-loaded NLCs (consisting of AOT, Tween 80, ethanol as the water phase; and Emulmetik 900, glyceryl monostearate, stearic acid, lexol as the oil phase) and simultaneously prepared polymer-coated NLCs (using polyvinyl alcohol, polyethylene glycol, and chitosan as the polymers). They showed that the polymer-coated NLCs exhibit greater mucoadhesion properties and physical stability than the uncoated NLCs.

3. Applications

Microemulsions as nanotemplates have engraved a prominent place despite the presence of innumerable methods for fabrication of NPs. Exploiting the microemulsion technique, distinct NPs, *viz.* metallic NPs, PNPs, QDs, MSN, SLNs, NLCs, etc., have been reported. The synthesized NPs have found applications in various fields like catalysis, delivery of drugs and diagnostics, sensing, etc. The recent trend of NPs derived from microemulsion method has been tabulated in **Table 1**.

S. no.	Type of nanoparticles	Microemulsion system (surfactant/ cosurfactant/oil/water)	Application of nanoparticles	Reference
1.	Ag	(n-Hexane/ethanol/water), silver nitrate as precursor and sodium borohydride as reducing agent	Catalysis	[67]
2.	Ni, Mo, Fe (transition nanoparticles)	(Sodium dodecyl-benzenesulfonate/citric acid/ toulene/1-hexanol), nickel (II) nitrate hexahydrate, iron(III) nitrate nonahydrate, ammonium molybdate tetrahydrate as precursor and sodium borohydride as reducing agent	*In situ* hydroprocessing of crude oil	[17]
3.	PMMA	Tween 80/ammonium persulfate/methyl methacrylate/quercetin hydrate	Drug delivery	[68]
4.	PAni and Ag/ PAni	Aniline/ammonium peroxydisulfate in W/O microemulsion(triton X 100/cyclohexane/1-butanol), silver nitrate, sodium borohydride as reducing agent	Antibacterial activity	[69]
5.	Silica-coated CdSe/ZnS QDs	Triton X-100/1-hexyl alcohol/cyclohexane/ hydrophobic CdSe/ZnS QDs	Sensing	[70, 71]
6.	MSN	CTAB/polydecene/cyclohexane and TEOS	Catalysis	[52]
7.	SLNs	Glyceryl monostearate/poloxamer 123	Catalysis	[60]
8.	SLNs	α-Tocopheryl linoleate/Tween 20/1-butanol/ biliary salt/α-linolenic acid/water	Drug Delivery	[72]
9.	NLCs	Stearic acid/castor oil/Imwitor 900/Tween 80/ sodium deoxy cholate	Drug delivery	[73]
10.	NLCs	Glyceryl monostearate/oleic acid/soy lecithin	Drug delivery	[74]

Table 1.
Recent overview of nanoparticles derived from the microemulsion method.

4. Conclusion

Microemulsions as nanotemplates have emerged as a soft and versatile approach for the fabrication of distinct nanoassemblies owing to their special ability to tune the particle properties such as the morphology, particle size, geometry, surface properties (activity and selectivity), etc. In this present chapter, a brief description of microemulsions as nanoreactors has been highlighted, stating the type of microemulsions (W/O or bicontinuous or O/W microemulsions) employed for generating different nanoparticles, including the mechanism involved behind the formation of nanoparticles using the microemulsion method. A detailed account of numerous nanoparticles such as metallic nanoparticles, quantum dots, polymeric nanoparticles, mesoporous silica nanoparticles, solid lipid nanoparticles, and nanostructured lipid carriers prepared from microemulsion method have been discussed, comprising their history, evolution, preparation, and applications. Although innumerable applications of the fabricated nanoassemblies have been reported, however, still the inbuilt potential of these exuberant nanocarriers has not been exploited completely.

Acknowledgements

Rohini Kanwar gratefully acknowledges CSIR and DAAD for fellowship. Jyoti is thankful to UGC for stipend. Madhuri Tanaji Patil is thankful to DST WOSA. Surinder Kumar Mehta acknowledges DST-PURSE II.

Conflict of interest

No conflict of interests is there to be declared.

Author details

Rohini Kanwar, Jyoti Rathee, Madhuri Tanaji Patil and Surinder Kumar Mehta*
Department of Chemistry and Centre for Advanced Studies in Chemistry, Panjab University, Chandigarh, India

*Address all correspondence to: skmehta@pu.ac.in

References

[1] Danielsson I, Lindman B. The definition of microemulsion. Colloids and Surfaces A. 1981;**3**:391-392. DOI: 10.1016/0166-6622(81)80064-9

[2] Hoar TP, Schulman JH. Transparent water-in-oil dispersions: The oleopathic hydro-micelle. Nature. 1943;**152**:102-103. DOI: 10.1038/152102a0

[3] Schulman JH, Stoeckenius W, Prince LM. Mechanism of formation and structure of micro emulsions by electron microscopy. The Journal of Physical Chemistry. 1959;**63**:1677-1680. DOI: 10.1021/j150580a027

[4] Lee KL. Applications and use of microemulsions. Imperial College London. 2010:6. DOI: 10.1007/978-1-4614-1205-2

[5] Mumper RJ, Cui Z, Oyewumi MO. Nanotemplate engineering of cell specific nanoparticles nanotemplate engineering of cell specific nanoparticles. Journal of Dispersion Science and Technology. 2003;**24**: 569-588. DOI: 10.1081/DIS-120021814

[6] Bowcott JE, Schulman JH. Emulsions. Zeitscrift für Elektrochemie. 1955;**59**: 283-290. DOI: 10.1002/bbpc.19550590409

[7] Robbins ML. Theory of the phase behavior of microemulsions. Society of Petroleum Engineers of AIME. 1976;**2**: 713ff

[8] Adamson AW. A model for micellar emulsions. Journal of Colloid and Interface Science. 1969;**29**:261-267. DOI: 10.1016/0021-9797(69)90195-7

[9] Malik MA, Wani MY, Hashim MA. Microemulsion method: A novel route to synthesize organic and inorganic nanomaterials. Arabian Journal of Chemistry. 2012;**5**:397-417. DOI: 10.1016/j.arabjc.2010.09.027

[10] Dhand C, Neeraj Dwivedi XJL, Ying ANJ, Verma NK, Beuerman RW, Rajamani Lakshminarayanan SR. Methods and strategies for the synthesis of diverse nanoparticles and their applications: A comprehensive overview. RSC Advances.2015;**5**:52. DOI: 10.1039/C5RA19388E

[11] Sanchez-Dominguez M, Pemartin K, Boutonnet M. Preparation of inorganic nanoparticles in oil-in-water microemulsions: A soft and versatile approach. Current Opinion in Colloid and Interface Science. 2012;**17**:297-305. DOI: 10.1016/j.cocis.2012.06.007

[12] Boutonnet M, Kizling J, Stenius P, Maire G. The preparation of monodisperse colloidal metal particles from microemulsions. Colloids and Surfaces. 1982;**5**:209-225. DOI: 10.1016/0166-6622(82)80079-6

[13] Destrée C, Nagy JB. Mechanism of formation of inorganic and organic nanoparticles from microemulsions. Advances in Colloid and Interface Science. 2006;**123–126**:353-367. DOI: 10.1016/j.cis.2006.05.022

[14] Pileni MP, Lisiecki I. Nanometer metallic copper particle synthesis in reverse micelles. Colloids and Surfaces A: Physicochemical and Engineering Aspects. 1993;**80**:63-68. DOI: 10.1016/0927-7757(93)80224-3

[15] Tianimoghadam S, Salabat A. A microemulsion method for preparation of thiol-functionalized gold nanoparticles. Particuology. 2018;**37**: 33-36. DOI: 10.1016/j.partic.2017.05.007

[16] Perez-Coronado AM, Soares OSGP, Calvo L, Rodriguez JJ, Gilarranz MA, Pereira MFR. Catalytic reduction of

bromate over catalysts based on Pd nanoparticles synthesized via water-in-oil microemulsion. Applied Catalysis B: Environmental. 2018;**237**:206-213. DOI: 10.1016/j.apcatb.2018.05.077

[17] Lam-Maldonado M, Melo-Banda JA, Macias-Ferrer D, Portales-Martínez B, Dominguez JM, Silva-Rodrigo R, et al. Transition metal nanocatalysts by modified inverse microemulsion for the heavy crude oil upgrading at reservoir. Catalysis Today. 2018:0-1. DOI: 10.1016/j.cattod.2018.05.052

[18] Yamagishi A, Masui T, Watanabe F. Temperature-jump study of aquation equilibriums of cobalt(II) ion solubilized in reversed micelles. The Journal of Physical Chemistry. 1980;**84**:34-40. DOI: 10.1021/j100438a010

[19] Lopez-Quintela MA. Synthesis of nanomaterials in microemulsions: Formation mechanisms and growth control. Current Opinion in Colloid and Interface Science. 2003;**8**:137-144. DOI: 10.1016/S1359-0294

[20] Ge JP, Chen W, Liu LP, Li YD. Formation of disperse nanoparticles at the oil/water interface in normal microemulsions. Chemistry: A European Journal. 2006;**12**:6552-6558. DOI: 10.1002/chem.200600454

[21] Sanchez-Dominguez M, Boutonnet M, Solans C. A novel approach to metal and metal oxide nanoparticle synthesis: The oil-in-water microemulsion reaction method. Journal of Nanoparticle Research. 2009;**11**: 1823-1829. DOI: 10.1007/s11051-009-9660-8

[22] Li H, Liu J, Xie S, Qiao M, Dai W, Li H. Highly active Co-B amorphous alloy catalyst with uniform nanoparticles prepared in oil-in-water microemulsion. Journal of Catalysis. 2008;**259**:104-110. DOI: 10.1016/j.jcat.2008.07.015

[23] Pemartin-Biernath K, Vela-Gonzalez AV, Moreno-Trejo MB, Leyva-Porras C, Castaneda-Reyna IE, Juarez-Ramïrez I, et al. Synthesis of mixed Cu/Ce oxide nanoparticles by the oil-in-water microemulsion reaction method. Materials (Basel). 2016;**9**: 480-501. DOI: 10.3390/ma9060480

[24] Hu Z, Nourafkan E, Gao H, Wen D. Microemulsions stabilized by in-situ synthesized nanoparticles for enhanced oil recovery. Fuel. 2017;**210**:272-281. DOI: 10.1016/j.fuel.2017.08.004

[25] Rivera-Rangel RD, González-Muñoz MP, Avila-Rodriguez M, Razo-Lazcano TA, Solans C. Green synthesis of silver nanoparticles in oil-in-water microemulsion and nano-emulsion using geranium leaf aqueous extract as a reducing agent. Colloids and Surfaces A: Physicochemical and Engineering Aspects. 2018;**536**:60-67. DOI: 10.1016/j.colsurfa.2017.07.051

[26] Ghasemi Y, Peymani P, Afifi S. Quantum dot: Magic nanoparticle for imaging, detection and targeting. Acta Bio-medica dell'Ateneo Parmense. 2009; **80**:156-165. DOI: 19848055

[27] Lim SY, Shen W, Gao Z. Carbon quantum dots and their applications. Chemical Society Reviews. 2015;**44**: 362-381. DOI: 10.1039/c4cs00269e

[28] Shen J, Zhu Y, Yang X, Li C. Graphene quantum dots: Emergent nanolights for bioimaging, sensors, catalysis and photovoltaic devices. Chemical Communications. 2012;**48**: 3686-3699. DOI: 10.1039/c2cc00110a

[29] Pathak S. Quantum dot applications to neuroscience: New tools for probing neurons and glia. The Journal of Neuroscience. 2006;**26**:1893-1895. DOI: 10.1523/JNEUROSCI.3847-05.2006

[30] Shakur HR. A detailed study of physical properties of ZnS quantum dots

synthesized by reverse micelle method. Physica E. 2011;**44**:641-646. DOI: 10.1016/j.physe.2011.10.021

[31] Yang L, Zhu J, Xiao D. Microemulsion-mediated hydrothermal synthesis of ZnSe and Fe-doped ZnSe quantum dots with different luminescence characteristics. RSC Advances. 2012;**2**:8179-8188. DOI: 10.1039/c2ra21401f

[32] Prasad SS. Synthesis and characterization of CdS quantum dots by reverse micelles method. Pharma Chemica. 2013;**5**:1-4. DOI: 10.1109/ICANMEET.2013.6609227

[33] Chen R, Han B, Yang L, Yang Y, Xu Y, Mai Y. Controllable synthesis and characterization of CdS quantum dots by a microemulsion-mediated hydrothermal method. Journal of Luminescence. 2016;**172**:197-200. DOI: 10.1016/j.jlumin.2015.12.006

[34] Tarkas HS, Marathe DM, Mahajan MS, Muntaser F, Patil MB, Tak SR, et al. Synthesis of tin monosulfide (SnS) nanoparticles using surfactant free microemulsion (SFME) with the single microemulsion scheme. Materials Research Express. 2017;**4**:1-8. DOI: 10.1088/2053-1591/aa57de

[35] Xia H, Wang Q. Synthesis and characterization of conductive polyaniline nanoparticles through ultrasonic assisted inverse microemulsion polymerization. Journal of Nanoparticle Research. 2001;**3**: 401-411. DOI: 10.1023/A:1012564814745

[36] Nagavarma BVN, Yadav HKS, Ayaz A, Vasudha LS, Shivakumar HG. Different techniques for preparation of polymeric nanoparticles—A review. Asian Journal of Pharmaceutical and Clinical Research. 2012;**5**:16-23

[37] Vauthier C, Bouchemal K. Methods for the preparation and manufacture of polymeric nanoparticles. Pharmaceutical Research. 2009;**26**:1025-1058. DOI: 10.1007/s11095-008-9800-3

[38] Stoffer JO, Bone T. Polymerization in water in oil microemulsion systems containing methyl methacrylate. Journal of Dispersion Science and Technology. 1980;**1**:37-54. DOI: 10.1080/01932698008962160

[39] Ming W, Jones FN, Fu S. Synthesis of nanosize poly(methyl methacrylate) microlatexes with high polymer content by a modified microemulsion polymerization. Polymer Bulletin. 1998; **40**:749-756. DOI: 10.1007/s002890050318

[40] Gan LM, Ng SC, Ong SP, Chew CH, Koh LL. A light scattering study on the droplet size and interdroplet interaction in microemulsions of AOT-oil-water system. Colloid and Polymer Science. 1989;**267**:1087-1095. DOI: 10.1016/0021-9797(88)90261-5

[41] Selvan ST, Mani A, Athinarayanasamy K, Phani KLN, Pitchumani S. Synthesis of crystalline polyaniline. Materials Research Bulletin. 1995;**30**:699-705. DOI: 10.1016/0025-5408(95)00065-8

[42] Wang QI, Xia H, Zhang C. Preparation of polymer/inorganic nanoparticles composites. Journal of Applied Polymer Science. 2001;**80**: 1478-1488

[43] Xia H, Zhang C, Wang Q. Study on ultrasonic induced encapsulating emulsion polymerization in the presence of nanoparticles. Journal of Applied Polymer Science. 2001;**80**: 1130-1139. DOI: 10.1002/app.1196

[44] Munshi N, De Tapas K, Maitra A. Size modulation of polymeric nanoparticles under controlled dynamics of microemulsion droplets. Journal of Colloid and Interface Science. 1997;**190**:387-391. DOI: 10.1006/jcis.1997.4889

[45] Zhi J, Wang Y, Lu Y, Ma J, Luo G. In situ preparation of magnetic chitosan/Fe_3O_4 composite nanoparticles in tiny pools of water-in-oil microemulsion. Reactive and Functional Polymers. 2006;**66**: 1552-1558. DOI: 10.1016/j. reactfunctpolym.2006.05.006

[46] Ge J, Jacobson GB, Lobovkina T, Holmberg K, Zare RN. Sustained release of nucleic acids from polymeric nanoparticles using microemulsion precipitation in supercritical carbon dioxide. Chemical Communications. 2010;**46**:9034-9036. DOI: 10.1039/ c0cc04258g

[47] Ahmad N, Alam MA, Ahmad R, Naqvi AA, Ahmad FJ. Preparation and characterization of surface-modified PLGA-polymeric nanoparticles used to target treatment of intestinal cancer. Artificial Cells, Blood Substitutes, and Biotechnology. 2018;**46**:432-446. DOI: 10.1080/21691401.2017.1324466

[48] Mehmood A, Ghafar H, Yaqoob S, Gohar UF, Ahmad B. Mesoporous silica nanoparticles: A review. Journal of Developing Drugs. 2017;**06**:174-188. DOI: 10.4172/2329-6631.1000174

[49] Rahman IA, Padavettan V. Synthesis of silica nanoparticles by sol-gel: Size-dependent properties, surface modification, and applications in silica-polymer nanocomposites—A review. Journal of Nanomaterials. 2012;**2012**: 1-15. DOI: 10.1155/2012/132424

[50] Finnie KS, Bartlett JR, Barbé CJA, Kong L. Formation of silica nanoparticles in microemulsions. Langmuir. 2007;**23**:3017-3024. DOI: 10.1021/la0624283

[51] Kao KC, Mou CY. Pore-expanded mesoporous silica nanoparticles with alkanes/ethanol as pore expanding agent. Microporous and Mesoporous Materials. 2013;**169**:7-15. DOI: 10.1016/j. micromeso.2012.09.030

[52] Shang L, Shi R, Waterhouse GIN, Wu LZ, Tung C-H, Yin Y, et al. Nanocrystals@hollow mesoporous silica reverse-bumpy-ball structure nanoreactors by a versatile microemulsion-templated approach. Small Methods. 2018:1800105. DOI: 10.1002/smtd.201800105

[53] Montenegro L, Lai F, Offerta A, Sarpietro MG, Micicchè L, Maccioni AM, et al. From nanoemulsions to nanostructured lipid carriers: A relevant development in dermal delivery of drugs and cosmetics. Journal of Drug Delivery Science and Technology. 2016; **32**:100-112. DOI: 10.1016/j. jddst.2015.10.003

[54] Morel S, Rosa Gasco M, Cavalli R. Incorporation in lipospheres of [d-Trp-6]LHRH. International Journal of Pharmaceutics. 1994;**105**:R1-R3. DOI: 10.1016/0378-5173(94)90466-9

[55] Morel S, Ugazio E, Cavalli R, Gasco MR. Thymopentin in solid lipid nanoparticles. International Journal of Pharmaceutics. 1996;**132**:259-261

[56] Ugazio E, Cavalli R, Gasco MR. Incorporation of cyclosporin A in solid lipid nanoparticles (SLN). International Journal of Pharmaceutics. 2002;**241**: 341-344. DOI: 10.1016/S0378-5173(02) 00268-5

[57] Ma P, Dong X, Swadley CL, Gupte A, Leggas M, Ledebur HC, et al. Development of idarubicin and doxorubicin solid lipid nanoparticles to overcome Pgp-mediated multiple drug resistance in leukemia. Journal of Biomedical Nanotechnology. 2009;**5**: 151-161. DOI: 10.1166/jbn.2009.1021

[58] Kuo YC, Chung CY. Solid lipid nanoparticles comprising internal Compritol 888 ATO, tripalmitin and cacao butter for encapsulating and releasing stavudine, delavirdine and saquinavir. Colloids and Surfaces.

B, Biointerfaces. 2011;**88**:682-690. DOI: 10.1016/j.colsurfb.2011.07.060

[59] Mehrad B, Ravanfar R, Licker J, Regenstein JM, Abbaspourrad A. Enhancing the physicochemical stability of β-carotene solid lipid nanoparticle (SLNP) using whey protein isolate. Food Research International. 2018;**105**: 962-969. DOI: 10.1016/j.foodres.2017. 12.036

[60] Kanwar R, Bhar R, Mehta SK. Designed meso-macroporous silica framework impregnated with copper oxide nanoparticles for enhanced catalytic performance. ChemCatChem. 2018;**10**:1-10. DOI: 10.1002/cctc. 201701630

[61] Doktorovová S, Araújo J, Garcia ML, Rakovský E, Souto EB. Formulating fluticasone propionate in novel PEG-containing nanostructured lipid carriers (PEG-NLC). Colloids and Surfaces. B, Biointerfaces. 2010;**75**:538-542. DOI: 10.1016/j.colsurfb.2009.09.033

[62] Kuo YC, Chung JF. Physicochemical properties of nevirapine-loaded solid lipid nanoparticles and nanostructured lipid carriers. Colloids and Surfaces. B, Biointerfaces. 2011;**83**:299-306. DOI: 10.1016/j.colsurfb.2010.11.037

[63] Khurana S, Bedi PMS, Jain NK. Development of nanostructured lipid carriers for controlled delivery of mefenamic acid. International Journal of Biomedical Nanoscience and Nanotechnology. 2012;**1**:247-266. DOI: 10.1504/IJBNN.2012.051218

[64] Shao Z, Shao J, Tan B, Guan S, Liu Z, Zhao Z, et al. Targeted lung cancer therapy: Preparation and optimization of transferrin-decorated nanostructured lipid carriers as novel nanomedicine for co-delivery of anticancer drugs and DNA. International Journal of Nanomedicine. 2015;**10**:1223-1233. DOI: 10.2147/IJN.S77837

[65] Kanwar R, Kaur G, Mehta SK. Revealing the potential of Didodecyldimethylammonium bromide as efficient scaffold for fabrication of nano liquid crystalline structures. Chemistry and Physics of Lipids. 2016; **196**:61-68. DOI: 10.1016/j. chemphyslip.2016.02.006

[66] Chanburee S, Tiyaboonchai W. Mucoadhesive nanostructured lipid carriers (NLCs) as potential carriers for improving oral delivery of curcumin. Drug Development and Industrial Pharmacy. 2017;**43**: 432-440. DOI: 10.1080/03639045. 2016.1257020

[67] Zhao S, Duan L, Xiao C, Li L, Liao F. Single metal of silver nanoparticles in the microemulsion for recyclable catalysis of 4-nitrophenol reduction. Journal of Advances in Nanomaterials. 2017;**2**:31-40

[68] Kajbafvala A, Salabat A. A novel one-step microemulsion method for preparation of quercetin encapsulated poly(methyl methacrylate) nanoparticles. Iranian Polymer Journal. 2017;**26**:651-662. DOI: 10.1007/ s13726-017-0550-0

[69] Fatema UK, Rahman MM, Islam MR, Mollah MYA, Susan MABH. Nanocomposites of polyaniline with silver nanoparticles prepared using water in oil microemulsions as antibacterial agents. Macromolecular Symposia. 2018;**379**:1700031. DOI: 10.1002/masy.201700031

[70] Shen H, Wang H, Tang Z, Niu JZ, Lou S, Du Z, et al. High quality synthesis of monodisperse zinc-blende CdSe and CdSe/ZnS nanocrystals with a phosphine-free method. CrystEngComm. 2009;**11**:1733-1738. DOI: 10.1039/b909063k

[71] Lv Y, Li J, Wu R, Wang G, Wu M, Shen H, et al. Silica-encapsulated quantum dots for highly efficient and

stable fluorescence immunoassay of C-reactive protein. Biochemical Engineering Journal. 2018;**137**:344-351. DOI: 10.1016/j.bej.2018.06.016

[72] Cassano R, Mellace S, Marrelli M, Conforti F, Trombino S. α-Tocopheryl linolenate solid lipid nanoparticles for the encapsulation, protection, and release of the omega-3 polyunsaturated fatty acid: In vitro anti-melanoma activity evaluation. Colloids and Surfaces. B, Biointerfaces. 2017;**151**: 128-133. DOI: 10.1016/j.colsurfb.2016.11.043

[73] Albekery MA, Alharbi KT, Alarifi S, Ahmad D, Omer ME, Massadeh S, et al. Optimization of a nanostructured lipid carriers system for enhancing the biopharmaceutical properties of valsartan. Digest Journal of Nanomaterials and Biostructures. 2017; **12**:381-389

[74] Singh SK, Hidau MK, Gautam S, Gupta K, Singh KP, Singh SK, et al. Glycol chitosan functionalized asenapine nanostructured lipid carriers for targeted brain delivery: Pharmacokinetic and teratogenic assessment. International Journal of Biological Macromolecules. 2018;**108**: 1092-1100. DOI: 10.1016/j.ijbiomac.2017.11.031

Chapter 4

Microemulsions as Nanoreactors to Obtain Bimetallic Nanoparticles

Concha Tojo, David Buceta and M. Arturo López-Quintela

Abstract

Microemulsions are frequently used as nanoreactors for the synthesis of bimetallic nanoparticles. The ability to manipulate the metal distribution in bimetallic nanoparticles is essential for optimizing applications, and it requires a deeper understanding of how compartmentalization of reaction medium affects nanoparticle synthesis. A simulation model was developed to predict the atomic structure of bimetallic nanoparticles prepared via microemulsion in terms of metals employed and microemulsion composition. The model was successfully proved by comparing theoretical and experimental Au/Pt STEM profiles. On this basis, the model becomes a strong tool to further enhance our knowledge of the complex mechanisms governing reactions in microemulsions and its impact on final nanostructures. The purpose of this study is to perform a comprehensive kinetic analysis of coreduction of different couple of metals in the light of the interplay between three kinetic parameters: intermicellar exchange rate, chemical reduction rates of the two metals, and reactants concentration. The particular combination of these factors determines the reaction rate of each metal, which in turn determines the final metal arrangement.

Keywords: bimetallic nanoparticles, microemulsions, reduction rate, intermicellar exchange rate, nanocatalysts

1. Introduction

From the pioneering work of Boutonnet et al. [1], the synthesis of nanoparticles in microemulsions has been widely investigated with a variety of technical applications in catalysis [2–4], photonics [5], and energy conversion and storage devices [6–8]. The microemulsion route allows to control the size and composition of nanoparticles. A microemulsion consists of nanometer-sized water droplets dispersed in the oil phase and stabilized by a surfactant film. Reactants can be dissolved in the nano-sized water droplets or reverse micelles and can be exchanged between them by direct material transfer during an interdroplet collision [9]. The intermicellar exchange allows the reactants to be carried by the same droplet, so the chemical reaction can proceed inside the nanoreactor. Due to the space limitation inside the micelle, nucleation and growth of the particle are restricted, so it can result in the formation of size-controlled particles. In spite of the complexity of the reaction medium, microemulsion route has several advantages when compared to traditional methods. The first one is that nanoparticle size is directly controlled by the water/surfactant ratio, so narrow size distributions can be obtained. Another advantage is that surfactants around the nanoparticles can be removed with ease and nanoparticles can be prepared at room temperature. In addition, the confinement of reactants inside micelles induces important changes in reactant concentrations, which strongly affect the reaction rates. Finally, in relation to catalysis, nanoparticles obtained by the microemulsion route present an improved catalytic behavior than particles with the same composition which are synthesized by traditional procedures [10, 11].

A variety of nanomaterials, ranging from metals [12–14], bimetallic structures [15–17], other inorganic nanoparticles [18–20], and organic compounds [21, 22], has been prepared by this approach. In the field of catalysis, microemulsion approach was successfully used to prepare different nanostructured catalytic materials [2, 10, 17, 23–25].

Nevertheless, microemulsion route present a challenge due to the difficulty in managing the material intermicellar exchange. As mentioned above, reactants are distributed in separate nanoreactors, so the whole process (chemical reaction, nucleation, and subsequent growth to build up final particles) is conditioned by the material exchange between them. This exchange is mainly dictated by the surfactant, which is located on the interface between water and oil phases. The hydrophilic portion of the surfactant is anchored into water and the lipophilic one into oil, forming a film which surrounds the micelle surface. It is believed that, when a micelle-micelle collision is violent enough, the surfactant film breaks up, allowing the material exchange. As a consequence, the rate of intermicellar exchange controls the reactants encounter and therefore plays a key role in chemical kinetics in microemulsions. The ease with which intermicellar channels are established as well as their size and stability are determined by the microemulsion composition, which in turn has been shown to affect final nanoparticle properties [26–28].

In the paper at hand, we are focused on the study of Pt/M (M = Au, Rh) nanoparticles synthesized in microemulsions. Platinum-based nanoparticles (NPs) exhibit remarkable electrocatalytic activity in many important chemical and electrochemical reactions including oxygen reduction reaction (ORR) and direct methanol oxidation [29]. Apart from the inherent chemical and physical properties of the constitutive metals, the catalytic activity, which is one of the more relevant applications of bimetallic nanoparticles, relies notably on the metal distribution, that is, on the intraparticle nanoarrangement [30]. Bimetallic nanoparticles can show four main mixing patterns: (a) core-shell structures, in which one metal forms the core and the second metal covers the first one forming the surrounding shell; (b) mixed structures, which are often called alloys; (c) multilayer structures [31]; and (d) sub-cluster segregated structures, characterized by a small number of heteroatomic bonds [12]. So, the control of bimetallic intrastructure, mainly within the first atomic layers from the surface [25, 32], is key for performance enhancement of bimetallic catalysts. Furthermore, the optimal metal distribution depends on the particular chemical reaction. Au-core/Pt-shell nanocatalyst exhibits an improved activity to catalyze formic acid electro-oxidation [33] or oxygen reduction reaction [34, 35]. On the contrary, an alloyed Pt-Au is better for electro-oxidation of methanol [36]. Therefore, an in-depth study aimed at tailoring well-defined structures will be of great interest.

Although the simultaneous reduction of the two metals by the microemulsion route is one of the most common procedures to control the size and composition of bimetallic nanoparticles [24, 37], the prediction of the resulting metal arrangement is complicated, as far as the current state-of-the-art is concerned. As a matter of fact, many studies designed to produce new nanoarrangements via microemulsions come from trial-and-error experiments, mainly due to the high number of involved synthetic variables and to their interaction with the inherent complexity of the reaction media. A robust tool for elucidating the interplay between the different factors concerning final bimetallic nanoarrangements is computer simulation. With the aim of understanding the different factors affecting final nanostructures, we perform a comprehensive kinetic analysis of coreduction of different couple of metals in the light of the interplay between three kinetic parameters: intermicellar exchange rate, chemical reduction rates of the two metals, and reactants concentration. The particular combination of these factors determines the reaction rate of each metal, which in turn defines the final metal arrangement.

2. The model

A model was developed to simulate the kinetic course of the two chemical reductions (see Ref. [38] for details). The reaction medium is a microemulsion, which is described as a set of micelles. The one-pot method is reproduced by mixing equal volumes of three microemulsions, each of which contains one of the three reactants (two metal precursors and the reducing agent R). This pattern of mixing reactants recreates the one-pot method, by which the two metal salts are simultaneously reduced.

2.1 Initial reactants concentration

Reactants are initially distributed throughout micelles using a Poisson distribution, that is, the occupation of all micelles is not similar. In this study, we present results using different values of metal precursors concentration, but keeping a proportion 1:1 of the two metals: $\langle cAuCl_4^- \rangle = \langle cPtCl_6^{2-} \rangle = \langle c \rangle = 2, 16, 32$, and 64 metal precursors in each micelle, which corresponds to 0.01, 0.08, 0.16, and 0.40 M, respectively, in a micelle with a radius of 4 nm. Au and Rh precursors ($AuCl_4^-$ and $RhCl_6^{3-}$) are represented by M^+. Calculations have been made under isolation conditions, that is, reducing agent R is in excess: $(\langle cR \rangle = 10\langle cPtCl_6^{2-} \rangle)$.

2.2 Microemulsion dynamics and time unit

Micelles move and collide with each other. The intermicellar collision is a key feature in kinetics in microemulsions, because upon collision micelles are able to establish a water channel, which allows the exchange of their contents (metal precursors, reducing agent, metallic atoms, and/or growing particles). The material intermicellar exchange makes possible the reactant encounter inside micelles and, as a consequence, it is determinant of chemical reactions to occur. The intermicellar collision is simulated by choosing a 10% of micelles at random. These selected micelles collide, fuse (allowing material intermicellar exchange), and then redisperse. One Monte Carlo step begins in each intermicellar collision and ends when the quantity of species carried by colliding micelles is revised in agreement to the exchange criteria described below.

2.3 Metal characterization: reduction rate ratio

The reduction rate of a metal A (v_A) can be related to the standard potential (ε^0_A) by means of the Volmer equation:

$$\frac{j_A}{j_B} = \frac{v_A n_A F}{v_B n_B F} = \frac{n_A F k_{red,A}\, c_{O,A} \exp\left[-\dfrac{\beta_A n_A F \varepsilon_A}{RT}\right]}{n_B F k_{red,B}\, c_{O,B} \exp\left[-\dfrac{\beta_B n_B F \varepsilon_B}{RT}\right]} = \exp\left[\frac{\beta nF(\varepsilon_B - \varepsilon_A)}{RT}\right] \quad (1)$$

where j_A is the current density, n_A is the number of electrons, F is the Faraday constant, $k_{red,A}$ is the chemical rate constant, β_A is the transfer coefficient, $c_{O,A}$ is the concentration of oxidized A, R is the gas constant, and T is temperature. When two metals A and B, initially at the same concentration ($c_{O,A} = c_{O,B}$), are reduced simultaneously to synthesize an A/B bimetallic nanoparticle, this equation can be simplified by assuming the following approximations: the number of electrons ($n_A = n_B = n$), the transfer coefficients ($\beta_A = \beta_A = \beta$), and the chemical rate constants ($k_{red,A} = k_{red,B} = k_{red}$) are equal. (One must keep in mind that main factor governing reduction rates is by electrochemical potential.) Under this condition, a simple relation between the rates of electron transfer of two species A and B and their standard potentials can be deduced.

$$\log \frac{v_A}{v_B} = \frac{1}{2.3} \frac{\beta n F(\varepsilon_B - \varepsilon_A)}{RT} \tag{2}$$

This equation supports the rule according to which the higher the difference between the standard potentials of the two metals, the higher the ratio between both reduction rates is.

2.3.1 Au/Pt nanoparticles

On the basis of Eq. (2), to simulate the reduction rate of Au/Pt nanoparticles, the standard reduction potential must be taken into account. When the Au precursor is $AuCl_4^-$, the standard reduction potential is $\varepsilon^0(AuCl_4^-) = 0.926$ V, which is higher than that of Pt precursor $\varepsilon^0(PtCl_6^{2-} = 0.742$ V). This results in a faster formation rate of Au particles. In fact, Au is reduced so quickly that kinetics cannot be studied by conventional methods, so stopped flow techniques were needed [39]. The color change occurs instantaneously, so Au reduction was simulated as fast as possible, that is, 100% of Au precursors located in colliding micelles react to produce Au atoms, whenever the amount of reducing agent was enough. The reduction rate parameter of a metal A (v_A) is the percentage of reactants inside colliding micelles which are reduced during a collision to give rise to products (A atoms). Regarding to Pt, its reduction rate was successfully simulated by using $v_{Pt} = 10\%$, that is, only a 10% of Pt precursor reacts in each collision ($v_{Pt} = 10\%$) [40]. In this way, Au/Pt nanoparticle formation is simulated by a reduction rate ratio $v_{Au}/v_{Pt} = 100/10 = 10$, that is, Au reduction is 10 times faster than Pt.

The two reductions can take place simultaneously within the same micelle. The metal precursors and/or reducing agent that did not react remain behind in the micelle and will be exchanged or react later.

2.3.2 Pt/Rh nanoparticles

In order to research the influence of another metal in the pair Pt/M on Pt reduction, a metal whose reduction rate would be 10 times slower than Pt was chosen. In this manner, the reduction rate ratio is the same as used to simulate Au/Pt nanoparticles, so the possible differences in the kinetic behavior and the final metal distributions cannot be supported by the difference between the standard potentials. Therefore, the reduction rate of Pt is the same as that of Au/Pt pair ($v_{Pt} = 10\%$), but now Pt is the faster metal. Taking into account the standard reduction potential of $RhCl_6^{3-}$, $\varepsilon^0(RhCl_6^{3-}) = 0.44$ V, this Rh precursor is a good candidate to be simulated as $v_{Rh} = 1\%$ (only a 1% of $RhCl_6^{3-}$ located in the colliding micelles will be reduced ($v_{Pt}/v_{Rh} = 10/1 = 10$)).

The number of each species located within each micelle is adjusted at each step in agreement with the possibility of chemical reduction and the intermicellar exchange criteria (see below). As the metallic atoms are produced in each micelle, they are assumed to be deposited on nanoparticle seed. That is, unlike for reactants, which are isolated within the micelle, all metal atoms inside a micelle are aggregated forming a growing nanoparticle. In order to calculate the metal distribution in the final bimetallic nanoparticle, the sequence of metals which are reduced is monitored in each micelle as a function of time.

2.4 Microemulsion characterization: intermicellar exchange criteria

Two different intermicellar exchange criteria are implemented depending on the nature of exchanged species. Metal precursor, reducing agent, and free metal atoms are isolated species, which will be redistributed between two colliding micelles in accordance with the concentration gradient principle: they are transferred from the more to the less occupied micelle. The exchange parameter k_{ex} quantifies the maximum amount of isolated species that can be exchanged during an intermicellar collision. As a result of this redistribution, the metal salts ($PtCl_6^{2-}$ and/or M^+) and the reducing agent R can be located within the same micelle. At this stage, chemical reduction can occur at a rate which depends on the nature of the metal.

As the reductions take place, metal atoms are produced within micelles. It is assumed that metal atoms are deposited on nanoparticle seed, so all metal atoms inside a micelle are considered to be aggregated forming a growing nanoparticle. The larger size of a growing nanoparticle leads to a second interdroplet exchange protocol. It is assumed that the exchange of growing particles is restricted by the size of the channel connecting colliding micelles. The ease with which this channel can be established as well as the channel size is mainly determined by the flexibility of the surfactant film. The flexibility parameter (f) specifies the maximum particle size for transfer between micelles. The exchange criterium of growing particles also takes into account Ostwald ripening, which assumes that larger particles grow by condensation of material, coming from the smaller ones that solubilize more readily than larger ones. This feature is included in the model by considering that if both colliding micelles carry a growing particle, the smaller one is exchanged towards the micelle carrying the larger one, whenever the channel size would be large enough.

As the synthesis advances, micelles can contain simultaneously reactants and growing particles. In this situation, autocatalysis can take place. Thus, if one of the colliding micelles is carrying a growing particle, the reaction always proceeds on it. If both colliding micelles contain particles, reaction takes place in the micelle containing the larger one, because it has a larger surface, so a higher probability of playing as catalyst.

Based on these simple criteria for material interdroplet exchange, surfactant film flexibility can be characterized as follows. There are two main requirements for material intermicellar exchange to occur: the size of the channel connecting colliding micelles must be large enough and the dimer formed by colliding micelles must be stable, that is, they must remain together long enough. Isolated species (reactants and free metals) traverse the intermicellar channel one by one, so one can assume that the key factor determining their exchange is the dimer stability. That is, when the two micelles stay together longer (higher dimer stability), a larger quantity of species can be exchanged. Channel size would not be relevant in this case. Based on this, k_{ex}, which quantifies how many units of isolated species can be exchanged during a collision, is related to the dimer stability. Conversely, when the transferred material is a particle constituted by aggregation of metal atoms, which travels through the channel as a whole, channel size becomes decisive. This kind of material exchange will be restricted by the intermicellar channel size (f parameter). From this picture, the flexibility of the surfactant film is simulated by means of these two parameters, k_{ex} (dimer stability) and f (intermicellar channel size).

A rigid film, such as AOT/n-heptane/water microemulsion, was successfully reproduced considering a channel size f = 5, associated to k_{ex} = 1 free atoms exchanged during a collision [26]. In case of flexible film, both factors rise together, because a more flexible film produces more stable dimer and larger channel size, allowing a quicker exchange of isolated species as well as an exchange of larger particles [41]. That is, a flexible film is associated to a faster material intermicellar exchange rate. A more flexible microemulsion, such as 75% Isooctane/20% Tergitol/5% water microemulsion, was successful compared to simulation data using the values f = 30, k_{ex} = 5 [42].

2.5 Description of the metal distribution in the bimetallic nanoparticle

The composition of each nanoparticle is revised at each step and monitored as a function of time. When all metal precursors were reduced and the content of all micelles remains constant over time, nanoparticle synthesis is considered to be finished. At this stage, the sequence of metal deposition of each particle (which is stored as a function of time) is stabilized. One simulation run produces a set of micelles, each one of them can carry one particle with different composition or be empty. At the end of each run, the averaged nanoparticle is calculated. Finally, results are averaged over 1000 runs.

The intrastructure of each particle is calculated by analyzing the sequence in which the two metals are deposited on the nanoparticle surface. So that, each sequence is arranged in 10 concentric layers, assuming that final nanoparticle is spherical. Then, the averaged percentage of each metal is calculated layer by layer. The final bimetallic distribution is represented by histograms, in which the layer composition is described by a color grading, as stated in the following pattern: Au, Pt, and Rh are represented by red, blue, and green, respectively. As the proportion of pure metal in the layer is higher, the color becomes lighter. In order to illustrate the heterogeneity of nanoparticle composition, the number of particles with a given percentage of the faster reduction metal (Au in Au/Pt and Pt in Pt/Rh nanoparticles) in each of 10 layers is also represented in the histograms. This analysis is reproduced layer by layer, from the beginning of the synthesis (inner layer or core) to the end (outer layer or surface). To simplify, the metal distribution is also shown by means of concentric spheres, whose thickness is proportional to the number of layers with the same composition, keeping the same color pattern.

3. Results and discussion

3.1 Factors affecting metal distribution: initial reactants concentration

The simulation model was successfully validated by comparison with experimental results. Au/Pt nanoparticles were synthesized in a 75% Isooctane/20% Tergitol/5% water microemulsion [42] (which can be characterized as flexible microemulsion)using different precursor concentrations. ($\langle cAuCl_4^- \rangle = \langle cPtCl_6^{2-} \rangle = \langle c \rangle$ = 0.01, 0.08, 0.16, and 0.40 M. The resulting Au/Pt particles were studied by HR-STEM and their structures were revealed by cross sections scanned with EDX analysis. The studied conditions were reproduced by simulation, using concentrations $\langle cAuCl_4^- \rangle = \langle cPtCl_6^{2-} \rangle = \langle c \rangle$ = 2, 16, 32, and 64 metal precursors in each micelle, which corresponds to 0.01, 0.08, 0.16, and 0.40 M, respectively, in a micelle with a radius of 4 nm. As mentioned above, Au/Pt pair was characterized as v_{Au}/v_{Pt} = 10 reduction rate ratio, and the 75% Isooctane/20% Tergitol/5% water microemulsion was simulated as a flexible surfactant film (f = 30, k_{ex} = 5).

The left column in **Figure 1** shows the simulated nanostructures obtained at each concentration. In order to compare the experimental and simulated nanostructures, the quantity of each metal crossed by a beam of 2 Å (approximate EDX beam size) was computed from each simulated final nanoparticle, and the theoretical STEM profiles were calculated. The STEM profiles of the average particle for each concentration are shown in center (theoretical) and right (experimental) columns of **Figure 1**. For a better comparison, experimental x-axis was changed from nm to counts, and the two kind of profiles were normalized to 1. Both profiles show the expected behavior: the surface (outer layers) is enriched in Pt, because of its slower reduction rate, and Au, which is reduced faster, accumulates in the core (inner layers). As concentration increases (see **Figure 1** from the top to the bottom), deeper Pt profiles are obtained. This means that the final nanostructure shows an improved metal segregation as concentration is higher. It is clearly observed in the histograms, which evolve from Au core covered by a mixed shell obtained at a low concentration to a more mixed Au core covered by a pure Pt shell as concentration

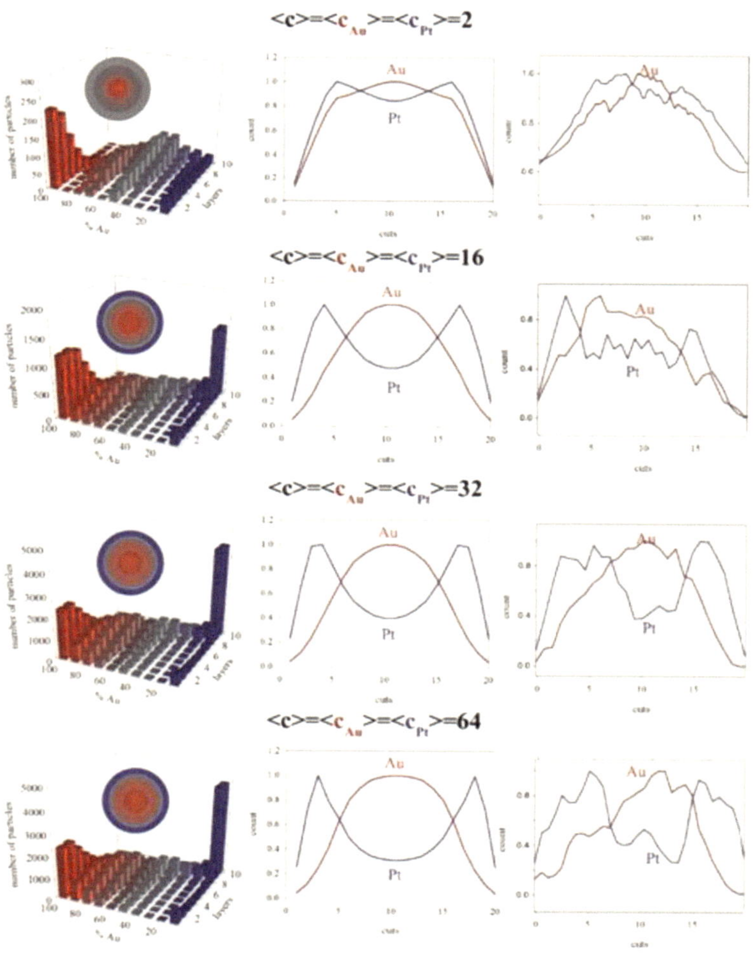

Figure 1.
Left column: simulated histograms for different initial concentrations (Au:Pt = 1:1). The proportion of pure metal in the layer is higher as the color becomes lighter (red: 100% Au, blue: 100% Pt, gray: 50% Au-Pt). Centre column: calculated STEM profiles for the average nanoparticle. Right column: measured STEM profiles for Au/Pt nanoparticles synthesized in a water/tergitol/isooctane microemulsion. Simulation parameters: flexible film (k_{ex} = 5, f = 30); reduction rate ratio (v_{Au}/v_{Pt} = 100/10); and reducing agent concentration $\langle cR \rangle = 10\langle M^{+} \rangle$. Adapted with permission from Ref. [42]. Copyright (2015) American Chemical Society.

increases (see decreasing red bar on the left and increasing blue bar on the right in histograms of **Figure 1**). This means that the nanostructure can be fine-tuned with sub-nanometer resolution, just by changing concentration.

A good agreement between experimental and theoretical results was attained, upholding the validity of the simulation model to predict the atomic structure of bimetallic nanoparticles. On this basis, the model becomes a strong tool to further enhance our knowledge of the complex mechanisms governing reactions in micro-emulsions and the impact of compartmentalization on final nanostructures.

The better metal segregation obtained as concentration increases is also observed when the two reductions are slowing down, as shown in **Figure 2**. This figure shows the final nanostructures obtained for the pair Pt/Rh (v_{Pt}/v_{Rh} = 10/1 = 10), under the same synthetic conditions as used in **Figure 1** to prepare Au/Pt nanoparticles. The better metal separation cannot be attributed to a larger reduction rate ratio, because in both cases the faster metal is 10 times faster than the slower one (v_{Au}/v_{Pt} = 100/10 = 10). The better metal separation obtained for Pt/Rh pair is more evident at low concentration, where an alloy is obtained for Au/Pt and a core-shell structure for Pt/Rh (compare histograms when $\langle c \rangle$= 2 in **Figures 1** and **2**). This means that, although the reduction rate ratio is similar, when the reduction rate of the faster metal slows down as in Pt, the other metal (Rh) is delayed even more. As a consequence, both reactions take place at different stages of the synthesis, resulting in better segregated structures.

3.2 Factors affecting metal distribution: reduction rate ratio

The difference between the standard potentials of the two metal precursors is believed to be the most relevant factor to determine the kinetics and the resulting bimetallic arrangement [43]. As established in Eq. (2), the higher the difference between the standard potentials of the two metals, the higher the ratio between both reduction rates is. It results in the earlier reduction of the faster reduction metal, which builds up the core and becomes the seed for the subsequent deposition of the slower metal, which forms the surrounding shell. On the contrary, when the two reduction rates are almost similar, a mixed nanoalloy is expected. In spite of this argumentation was initially proposed for reactions in homogeneous media, and it does not take into account the confinement of reactants within micelles, it is frequently applied to explain results in microemulsion. As a rule, it is observed a tendency from nanoalloy to core-shell structure as difference in reduction potential is increased (see Table 1 in Ref. [44]). Previous simulation studies allow to clearly observe a better separation of the two metals as reduction rate ratio is larger (for a deeper discussion, see Ref. [28]).

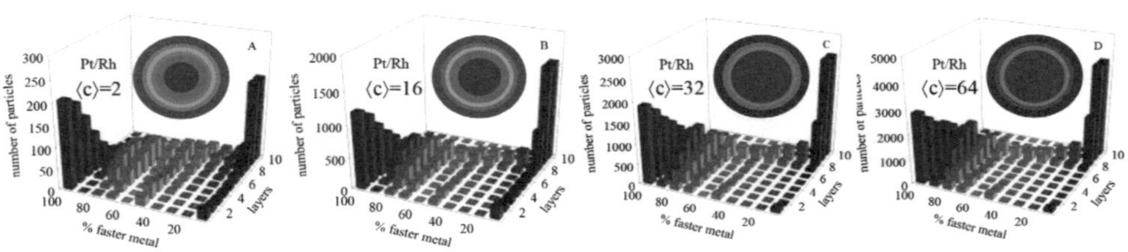

Figure 2.
Histograms show the number of particles with a given percentage of the faster reduction metal (Pt) in each layer at different concentrations. In all cases, $\langle cR \rangle = 10\langle cPtCl_6^{2-} \rangle$, *and* $cPtCl_6^{2-}:RhCl_6^{3-}$ *is in 1:1 proportion. Reduction rates:* v_{Pt} = 10%, v_{Rh} = 1% (v_{Pt}/v_{Rh} = 10); *flexible film* (f = 30, k_{ex} = 5). *Scheme color: Pt and Rh are represented by blue and green bars, respectively. Lighter colors mean less mixture. The nanostructure is also shown by colored concentric spheres, keeping the same color pattern.*

3.3 Factors affecting metal distribution: microemulsion composition

To isolate the effect of microemulsion composition on nanoparticle structure, a particular pair of metals must be chosen and analyze if a change in the microemulsion composition leads to a different metal segregation. For example, Pt/Ru nanoparticles were obtained as nanoalloy, both for rigid (water/Brij-30/n-heptane [45]) and flexible films (water/Berol 050/isooctane 80 [46] or water/NP5-NP9/cyclohexane [47]). But in this couple, the small difference in reduction potentials leads to quite similar reduction rates, which hinder metal segregation, even with a slow intermicellar exchange rate. As a matter of fact, when couples with higher reduction rate ratio are studied (such as Au/Ag, Au/Pt, and Au/Pd), an increase in surfactant flexibility results in the expected transition from a core-shell to a nanoalloy. As an example, alloyed Au/Pt nanoparticles were prepared using a flexible film such as water/Tergitol 15-S-5/isooctane [17] or water/TritonX-100/cyclohexane [48]. On the contrary, rigid films (water/AOT/isooctane [39] and water/Brij 30/n-heptane [49]) lead to segregated structures. The simulation model also predicts this result, as shown in **Figure 3**, in which different Au/Pt (v_{Au}/v_{Pt} = 10) arrangements were obtained by employing different values of surfactant film flexibilities. The ability of the microemulsion to minimize the difference between the reduction rates is clearly reflected in the progressive mixture of Au and Pt as increasing the intermicellar exchange rate (for a deeper discussion, see the following sections).

3.4 Kinetic study

The results shown in previous figures were obtained under isolation conditions, that is, the reducing agent concentration is much higher than stoichiometry, so the change in R concentration during the course of the reaction is negligible. As a result, the metal reduction, which is a bimolecular reaction, appears to be first order, when the reaction media is homogeneous. In order to study how the confinement of reactants inside micelles would affect chemical kinetics, the depletion of the number of metal precursors M^+ (M^+ = $AuCl_4^-$, $PtCl_6{}_-^{2-}$, $RhCl_6^{3-}$) was monitored as the synthesis advances. The logarithmic plot of M^+ concentration versus time is shown in **Figure 4** using different initial reactant concentrations. Left and right columns show results for Au/Pt and Pt/Rh couples, respectively. Au, Pt, and Rh are represented by dashed, solid, and dashed-dotted lines, respectively. **Figure 4A** and **B** was obtained by simulating a flexible film and **C** and **D** a rigid one. At first

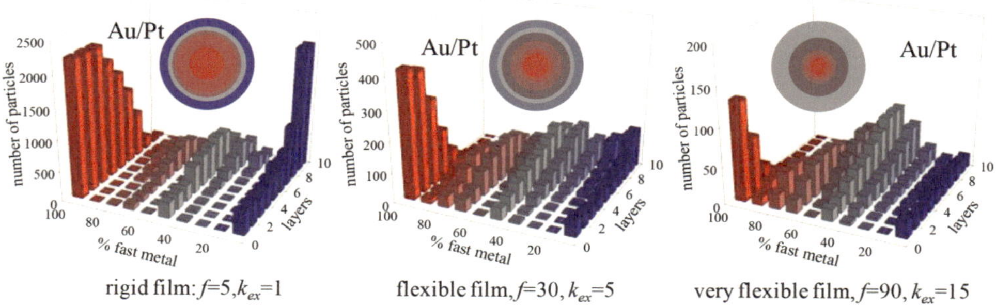

Figure 3.
Number of particles with a given percentage of the faster reduction metal (Au) in each layer using three different microemulsion compositions (different f and k_{ex} parameters). $\langle cAuCl_4^-\rangle$ = $\langle cPtCl_6^{2-}\rangle$ = 4; $\langle cR\rangle$ = 10$\langle cPtCl_6^{2-}\rangle$ reduction rates: v_{Au} = 100%, v_{Pt} = 10%. Scheme color: Au and Pt are represented by red and blue bars, respectively. Lighter colors mean less mixture. Adapted with permission from Ref. [44].

sight, metal reductions obey first-order kinetics in both Au/Pt and Pt/Rh synthesis, as expected. Nevertheless, it is important to note that, with the exception of Rh, a time lag is required to reach the linear regime. Two points must be highlighted: First, the higher the concentration, the longer the time lag between the beginning of the synthesis and the achievement of the linear behavior. On the second hand, the time lag strongly depends on the intermicellar exchange rate, being longer as the exchange rate is slower (rigid film). Both factors (concentration and film flexibility) suggest that the rate-determining step is the intermicellar exchange rate at earlier reaction times, as explained as follows. The synthesis starts when the microemulsions containing the reactants are mixed. In order to be able to react, reactants must be located inside the same micelle. The reactants redistribution between micelles is dictated by the rate with which reactants can go through the channels communicating colliding micelles, that is, the intermicellar exchange rate. So, a slow exchange rate only allows the exchange of few reactants in each collision, which implies that more collisions are required to redistribute reactants and allow the reactants encounter. Therefore, a rigid film requires much longer lag times than a flexible one (compare **Figure 4A** with **B** and **C** with **D**, for any value of concentration). Apart from that, reactants redistribution is also affected by concentration, because the number of reactants which can traverse the intermicellar channel during an effective collision is restricted. Therefore, if concentration within micelle is large, more collisions (i.e., more time) are needed to make possible reactants redistribution. Finally, it is interesting to point out that this delay in reaching linear behavior disappears when a very slow chemical reduction takes place, as shown by Rh kinetics in right column of **Figure 4** (see dashed-dotted lines), which is linear from the beginning, at any

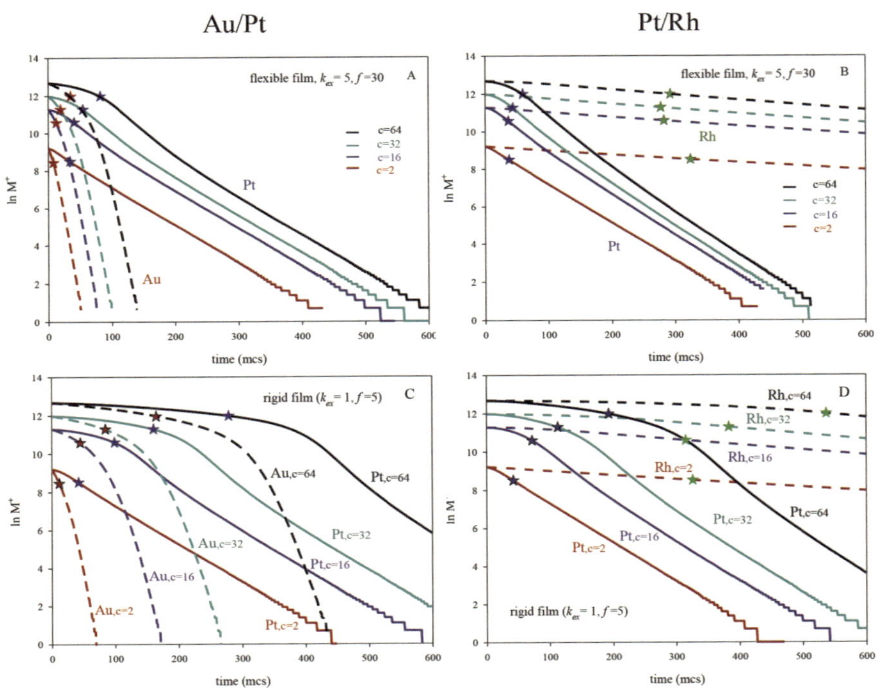

Figure 4.
Plot of ln M⁺ (number of metal salt) against time (in Monte Carlo step, mcs) using different initial concentration c (metal salts/micelle). Solid, dashed, and dashed-dotted represent Au, Pt, and Rh, respectively. A and B shows results for a flexible surfactant film (k_{ex} = 5, f = 30) and C and D for a rigid one (k_{ex} = 1, f = 5). Au/Pt pair (v_{Au}/v_{Pt} = 100/10 = 10) is represented in A and C and Pt/Rh pair (v_{Pt}/v_{Rh} = 10/1 = 10) in B and D. Stars indicate the half-life (red, blue, and green means Au, Pt and Rh, respectively).

concentration value. This behavior can be taken as indication that the reduction rate is so slow (only a 1% of reactants give rise to products) that the limiting step is the reduction itself.

Summarizing, the time lag required to achieve linear behavior in **Figure 4** reflects the time it takes for reactants to encounter. This time lag can be determinant of final metal segregation, because the inner layers of nanoparticle are building up during this stage.

Pseudo-first-order rate constants, k_{obs}, can be calculated from linear regime of the logarithmic plot as shown in **Figure 4**. The values of the pseudo-first-order rate constants are represented in **Figure 5** as a function of concentration. One can observe that the slopes of Au reduction are always higher than the slopes of Pt, which in turn is faster that Rh, as expected. Classical chemical kinetics in a homogeneous reaction medium establishes that k_{obs} in bimolecular reactions does not depend on precursors concentrations. This is the case for Pt and Rh, whose k_{obs} values did not depend neither on the concentration nor on the intermicellar exchange rate. In contrast, k_{obs} values of Au are strongly influenced by both factors. To explain this behavior, one have to take into account that the limiting step in Au chemical reduction is the intermicellar exchange [38, 50], because of the extremely fast Au reduction rate. One can observe that the dependence of k_{obs} on concentration decreases as intermicellar exchange rate is faster, until reaching an almost constant value at very fast intermicellar exchange rate (see gray line in **Figure 5**), as expected. In comparison, Pt and Rh reductions are so slow that their rates are not limited by the exchange rate.

One can conclude from **Figure 5** that intermicellar exchange rate exerts a different degree of influence depending on the reduction rate of the metal in comparison to the intermicellar exchange rate. This means that the compartmentalization of reaction medium takes part in chemical kinetics more or less depending on the metal nature. This different interplay between exchange rate and reduction rate has to be reflected in the metal segregation of final nanoparticle. It was proposed that if the intermicellar exchange rate can only modify the rate of metals whose

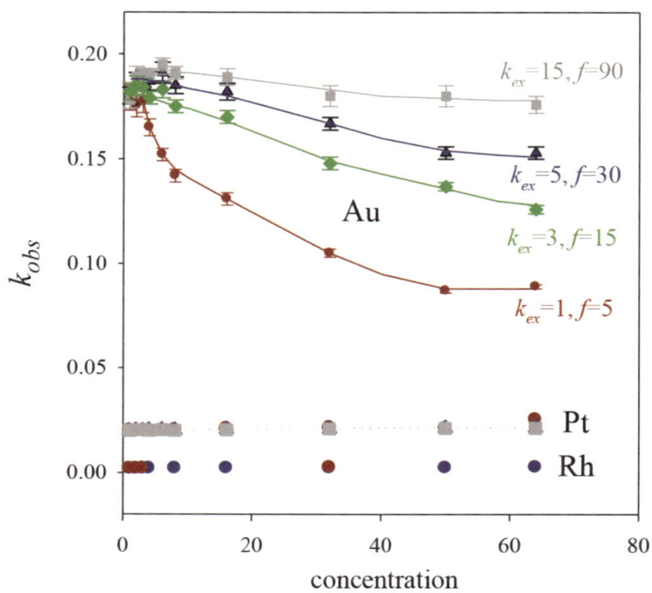

Figure 5.
*k_{obs} (slopes of the linear parts from the plots in **Figure** 3) as a function of concentration for different microemulsion compositions and different metals. v_{Au} = 100, v_{Pt} = 10, v_{Rh} = 1. Lines are only a guide to the eye.*

reduction is very fast [38], such as Au, only bimetallic nanoparticle including Au could be prepared with different metal distributions as a function of microemulsion composition (intermicellar exchange rate) by a one-pot method (see Table 1 in Ref. [44]). To the best of our knowledge, only Au/Pt, Au/Ag, and Au/Pd have been synthesized in a different intrastructure by different authors. Thus, when a rigid film (such as provided by AOT) is used, Au-Pt nanoparticles are arranged in a core-shell distribution [39]. On the contrary, more flexible surfactants such as Brij30 [49], tergitol [17], or TritonX-100 [48] give rise to alloyed nanoparticles. In relation to Au/Ag, alloys were obtained with TritonX-100 [51] and $C_{11}E_3$ and $C_{11}E_5$ [52], but an enriched in Ag surface was observed when microemulsion contains AOT [53]. Finally, AOT was also used to obtain core-shell Au-Pd nanoparticles [54] and alloys with Pd-rich surface [55]. In contrast, true Au-Pd alloys [4] have been obtained with Brij30 and TritonX-100. With that in mind, it could be suggested that metal segregation in the nanoparticle can be modified by a change in the microemulsion composition only when one of the metals is Au or another very fast reduction rate metal. Nevertheless, in spite of the agreement between theoretical and experimental data, this assumption is based on the kinetic constants, which were calculated from the linear plot shown in **Figure 4**. It must be emphasized that the linear regime is not fulfilled at initial stages of the reaction, when the core is been building up. With the aim of studying the relevance of the non-linear behavior at the beginning of the synthesis, the half-life, defined as the time that it takes for the reactant concentration to decay to half of its initial value, was calculated for each case. Stars in **Figure 4** show half-life under different synthesis conditions (Au, Pt, and Rh are represented by red, blue, and green stars, respectively). With the exception of Rh and Pt at very low concentration, half-life is usually smaller than the time needed to achieve the linear regime. As observed in **Figure 4C** and **D**, Au and Pt reductions in a rigid microemulsion and at high concentration have a half-life much earlier than linear plot. This means that not only the initial layers but also the middle ones are formed under a nonlinear regime. So, chemical reductions are still not a first-order process during the formation of a large part of the particle (for a deeper discussion on life time, see Ref. [38]).

4. Conclusions

The generalized belief according to which the difference in the reduction potentials determines the intrastructure in a bimetallic nanoparticle should be improved. We propose that there are three potentially limiting factors which restrict chemical kinetics of bimetallic nanoparticles prepared from microemulsions: chemical reduction rate itself, exchange rate of reactants between micelles, and reactants concentration. The specific combination of these three factors determines the reaction rate of each metal, which in turn determines the sequence of metals deposition and the resulting bimetallic arrangement.

The kinetic study of Pt/M nanoparticles prepared via microemulsions under isolation conditions shows that chemical reductions are pseudo-first-order reactions, but not from the initial stages. At the beginning of the synthesis, the reactants encounter is dictated by the redistribution of reactants between micelles, which is controlled by the intermicellar exchange rate. As a result, the limiting step of faster reduction metals, such as Au, is the intermicellar exchange. On the contrary, microemulsion dynamics has a little effect if reduction rates are very slow (i.e., Rh). This means that compartmentalization of the reaction media has a different impact depending on the reduction rate of the particular metal. We are not referring only to the reduction rate of a metal in relation to the another metal in the pair but also how fast the reduction takes place in relation to the intermicellar exchange rate. Specifically, for a given reduction rates ratio and keeping fixed microemulsion composition and concentration, the fact that the two reactions were slow (as in Pt/Rh) leads to a better metal segregation than if both reactions are faster (as in Au/Pt). Therefore, with the exception of very slow metal reduction as Rh, intermicellar exchange rate drastically impacts on chemical kinetics, particularly at the beginning of the synthesis. This is not a minor matter, because it will be reflected in the composition of the core and middle layers of the resulting nanoparticle. So, the ability of microemulsion to manipulate the sequence of metal deposition, when the metal reductions are quite fast in relation to the intermicellar exchange rate, can be used to design the experiments to synthesize bimetallic particles with ad hoc nanoarrangements. This ability disappears when the two chemical reductions are slow, because of chemically controlled kinetics.

In this paper, computer simulation has proved to be very useful tool to identify suitable synthesis parameters, which control metal segregation in a bimetallic nanoparticle. Further insights into the interplay between metal nature, exchange rate, and final bimetallic structure can be gained from kinetics studies.

Acknowledgements

This work was supported by MINECO, Spain (MAT2015-67458-P, co-financed with FERDER Funds), from the European Union's H2020 research and innovation programme under grant agreement No. 646155 (INSPIRED) and from Xunta de Galicia (Programa REDES ED431D-2017/18). D.B. thanks for the postdoc grant from Xunta de Galicia, Spain (POS-A/2013/018).

Conflict of interest

The authors declare no conflict of interest.

Notes/thanks/other declarations

This research is dedicated to Prof. Julio Casado Linarejos, who teached us the fundamentals of chemical kinetics.

Author details

Concha Tojo[1*], David Buceta[2] and M. Arturo López-Quintela[2]

1 Physical Chemistry Department, University of Vigo, Vigo, Spain

2 Laboratorio de Magnetismo y Nanotecnología, University of Santiago de Compostela, Santiago de Compostela, Spain

*Address all correspondence to: ctojo@uvigo.es

References

[1] Boutonnet M, Kizling J, Stenius P, Maire G. The preparation of monodisperse colloidal metal particles from microemulsions. Colloids and Surfaces. 1982;**5**:209-225

[2] Sánchez-Dominguez M, Boutonnet M. Synthesis of nanostructured catalytic materials from microemulsions. Catalysts. 2016;**6**:4-8

[3] Heshmatpour F, Abazari R. Formation of dispersed palladium-nickel bimetallic nanoparticles in microemulsions: Synthesis, characterization, and their use as efficient heterogeneous recyclable catalysts for the amination reactions of aryl chlorides under mild conditions. RSC Advances. 2014;**4**:55815-55826

[4] Li T, Zhou H, Huang J, Yin J, Chen Z, et al. Facile preparation of Pd-Au bimetallic nanoparticles via in-situ self-assembly in reverse microemulsion and their electrocatalytic properties. Colloids and Surfaces, A: Physicochemical and Engineering Aspects. 2014;**463**:55-62

[5] Yuan P, Ma R, Gao N, Garai M, Xu Q-H. Plasmon coupling-enhanced two-photon photoluminescence of Au@Ag core-shell nanoparticles and applications in the nuclease assay. Nanoscale Research Letters. 2015;7:10233-10239

[6] Gao M-R, Xu Y-F, Jiang J, Yu S-H. Nanostructured metal chalcogenides: Synthesis, modification, and applications in energy conversion and storage devices. Chemical Society Reviews. 2013;**42**:2986-3017

[7] Tuaev X, Rudi S, Petkov V, Hoell A, Strasser P. In situ study of atomic structure transformations of Pt-Ni nanoparticle catalysts during electrochemical potential cycling. ACS Nano. 2013;7:5666-5674

[8] Gu X, Lu Z-H, Jiang H-L, Akita T, Xu Q. Synergistic catalysis of metal-organic framework-immobilized Au-Pd nanoparticles in dehydrogenation of formic acid for chemical hydrogen storage. Journal of the American Chemical Society. 2011;**133**:11822-11825

[9] Fletcher PDI, Howe AM, Robinson BH. The kinetics of solubilisate exchange between water droplets of a water-in-oil microemulsion. Journal of the Chemical Society, Faraday Transactions. 1987;**83**:985-1006

[10] Eriksson S, Nylen U, Rojas S, Boutonnet M. Preparation of catalysts from microemulsions and their applications in heterogeneous catalysis. Applied Catalysis A. 2004;**265**:207-219

[11] Boutonnet M, Lögdberg S, Svensson EE. Recent developments in the application of nanoparticles prepared from w/o microemulsions in heterogeneous catalysis. Current Opinion in Colloid & Interface Science. 2008;**13**:270-286

[12] Magno LMA, Angelescu DG, Sigle W, Stubenrauch C. Microemulsions as reaction media for the synthesis of Pt nanoparticles. Physical Chemistry Chemical Physics. 2011;**13**:3048-3058

[13] Eastoe J, Hollamby MJ, Hudson LK. Recent advances in nanoparticle synthesis with reversed micelles. Advances in Colloid and Interface Science. 2006;**128-130**:5-15

[14] Yashima M, Falk LKL, Palmqvist AEC, Holmberg K. Structure and catalytic properties of nanosized alumina supported platinum and palladium particles synthesized by reaction in microemulsion. Journal of Colloid and Interface Science. 2003;**268**:348-356

[15] Szumelda T, Drelinkiewicz A, Kosydar R, Góral-Kurbiel M, Gurgul J, Duraczynska D. Formation of Pd-group VIII bimetallic nanoparticles by the "water-in-oil" microemulsion method. Colloids and Surfaces, A: Physicochemical and Engineering Aspects. 2017;**529**:246-260

[16] Magno LM, Sigle W, van Aken PA, Angelescu D, Stubenrauch C. Size control of PtPb intermetallic nanoparticles prepared via microemulsions. Physical Chemistry Chemical Physics. 2011;**20**:9134-9136

[17] Hernández-Fernández P, Rojas S, Ocón P, Gómez de la Fuente JL, San Fabián J, et al. Influence of the preparation route of bimetallic Pt-Au nanoparticle electrocatalyst for the oxygen reduction reaction. The Journal of Physical Chemistry. B. 2007;**111**:2913-2923

[18] Summers M, Eastoe J, Davis S. Formation of $BaSO_4$ nanoparticles in microemulsions with polymerized surfactant shells. Langmuir. 2002;**18**:5023-5026

[19] Holmberg K. Surfactant-templated nanomaterials synthesis. Journal of Colloid and Interface Science. 2004;**274**:355-364

[20] Sánchez-Domínguez M, Aubery C, Solans C. New trends on the synthesis of inorganic nanoparticles using microemulsions as confined reaction media. In: Hashim A, editor. Smart Nanoparticles Technology. Rijeka: Intech; 2012

[21] Aubry J-M, Bouttemy S. Preparative oxidation of organic compounds in microemulsions with singlet oxygen generated chemically by the sodium molybdate/hydrogen peroxide system. Journal of the American Chemical Society. 1997;**119**:5286-5294

[22] Holmberg K. Organic reactions in microemulsions. European Journal of Organic Chemistry. 2007;**5**:723-728

[23] Parapat RY, Wijaya M, Schwarze M, Selve S, Willinger M, Schomäcker R. Particle shape optimization by changing from an isotropic to an anisotropic nanostructure: Preparation of highly active and stable supported Pt catalysts in microemulsions. Nanoscale Research Letters. 2013;**5**:796-805

[24] Parapat RY, Parwoto V, Schwarze M, Zhang B, Su D-S, Schomäcker R. A new method to synthesize very active and stable supported metal Pt catalysts: Thermo-destabilization of microemulsions. Journal of Materials Chemistry. 2012;**22**:11605-11614

[25] König RYG, Schwarze M, Schomäcker R, Stubenrauch C. Catalytic activity of mono- and bi-metallic nanoparticles synthesized via microemulsions. Catalysts. 2014;**4**:256-275

[26] Tojo C, Blanco MC, López-Quintela MA. Preparation of nanoparticles in microemulsions: A Monte Carlo study of the influence of the synthesis variables. Langmuir. 1997;**13**:4527-4534

[27] Magno LM, Sigle W, Aken PAV, Angelescu DG, Stubenrauch C. Microemulsions as reaction media for the synthesis of bimetallic nanoparticles: Size and composition of particles. Chemistry of Materials. 2010;**22**:6263-6271

[28] Tojo C, de Dios M, López-Quintela MA. On the structure of bimetallic nanoparticles synthesized in microemulsions. Journal of Physical Chemistry C. 2009;**113**:19145-19154

[29] Stamenkovic VR, Mun BS, Arenz M, Mayrhofer KJ, Lucas CA, et al. Trends in electrocatalysis on extended and nanoscale Pt-bimetallic alloy surfaces. Nature Materials. 2007;**6**:241-247

[30] Shi J. On the synergetic catalytic effect in heterogeneous nanocomposite catalysts. Chemical Reviews. 2013;**113**:2139-2181

[31] Ferrando R, Jellinek J, Johnston RL. Nanoalloys: From theory to applications of alloy clusters and nanoparticles. Chemical Reviews. 2008;**108**:845-910

[32] Spanos I, Dideriksen K, Kirkensgaard JJK, Jelavic S, Arenz M. Structural disordering of de-alloyed Pt bimetallic nanocatalysts: The effect on oxygen reduction reaction activity and stability. Physical Chemistry Chemical Physics. 2015;**17**:28044-28053

[33] Zhang G-R, Zhao D, Feng Y-Y, Zhang B, Su DS, et al. Catalytic Pt-on-au nanostructures: Why Pt becomes more active on smaller au particles. ACS Nano. 2012;**6**:2226-2236

[34] Shao M, Peles A, Shoemaker K, Gummalla M, Njoki PN, et al. Enhanced oxygen reduction activity of platinum monolayer on gold nanoparticles. Journal of Physical Chemistry Letters. 2011;**2**:67-72

[35] Notar Francesco I, Fontaine-Vive F, Antoniotti S. Synergy in the catalytic activity of bimetallic nanoparticles and new synthetic methods for the preparation of fine chemicals. ChemCatChem. 2014;**6**:2784-2791

[36] Zhao L, Thomas JP, Heinig NF, Abd-Ellah M, Wang X, Leung KT. Au-Pt alloy nanocatalysts for electro-oxidation of methanol and their application for fast-response non-enzymatic alcohol sensing. Journal of Materials Chemistry C. 2014;**2**:2707-2714

[37] Yin Z, Ma D, Bao X. Emulsion-assisted synthesis of monodisperse binary metal nanoparticles. Chemical Communications. 2010;**46**:1344-1346

[38] Tojo C, Buceta D, López-Quintela MA. Bimetallic nanoparticles synthesized in microemulsions: A computer simulation study on relationship between kinetics and metal segregation. Journal of Colloid and Interface Science. 2018;**510**:152-161

[39] Wu M, Chen D, Huang T. Preparation of Au/Pt bimetallic nanoparticles in water-in-oil microemulsions. Chemistry of Materials. 2001;**13**:599-606

[40] Tojo C, Buceta D, López-Quintela MA. Understanding the metal distribution in core-shell nanoparticles prepared in micellar media. Nanoscale Research Letters. 2015;**10**:339-349

[41] Quintillán S, Tojo C, Blanco MC, López-Quintela MA. Effects of the intermicellar exchange on the size control of nanoparticles synthesized in microemulsions. Langmuir. 2001;**17**:7251-7254

[42] Buceta D, Tojo C, Vukmirovik M, Deepak FL, López-Quintela MA. Controlling bimetallic nanostructures by the microemulsion method with sub-nanometer resolution using a prediction model. Langmuir. 2015;**31**:7435-7439

[43] Feng J, Zhang C. Preparation of Cu-Ni alloy nanocrystallites in water-in-oil microemulsions. Journal of Colloid and Interface Science. 2006;**293**:414-420

[44] Tojo C, Buceta D, López-Quintela MA. On metal segregation of bimetallic nanocatalysts prepared by a one-pot method in microemulsions. Catalysts. 2017;**7**:1-17

[45] Solla-Gullón J, Vidal-Iglesias FJ, Montiel V, Aldaz A. Electrochemical characterization of platinum-ruthenium nanoparticles prepared by water-in-oil microemulsion. Electrochimica Acta. 2004;**49**:5079-5088

[46] Rojas S, García-García FJ, Jaeras S, Martínez-Huerta MV, García Fierro JL, Boutonnet M. Preparation of carbon supported Pt and PtRu nanoparticles from microemulsion. Applied Catalysis, A: General. 2005;**285**:24-35

[47] Liu Z, Lee JY, Han M, Chen W, Gan LM. Synthesis and characterization of PtRu/C catalysts from microemulsions and emulsions. Journal of Materials Chemistry. 2002;**12**:2453-2458

[48] Pal A. Gold–platinum alloy nanoparticles through water-in-oil microemulsion. Journal of Nanostructure in Chemistry. 2015;**5**:65-69

[49] Habrioux A, Vogel W, Guinel M, Guetaz L, Servat K, et al. Structural and electrochemical studies of Au-Pt nanoalloys. Physical Chemistry Chemical Physics. 2009;**11**:3573-3579

[50] Tojo C, de Dios M, Buceta D, López-Quintela MA. Cage-like effect in Au-Pt nanoparticle synthesis in microemulsions: A simulation study. Physical Chemistry Chemical Physics. 2014;**16**:19720-19731

[51] Pal A, Shah S, Devi S. Preparation of silver, gold and silver-gold bimetallic nanoparticles in w/o microemulsion containing triton X-100. Colloids and Surfaces, A: Physicochemical and Engineering Aspects. 2007;**302**:483-487

[52] Cheng J, Bordes R, Olsson E, Holmberg K. One-pot synthesis of porous gold nanoparticles by preparation of Ag/Au nanoparticles followed by dealloying. Colloids and Surfaces, A: Physicochemical and Engineering Aspects. 2013;**436**:823-829

[53] Chen D, Chen C. Formation and characterization of Au-Ag bimetallic nanoparticles in water-in-oil microemulsions. Journal of Materials Chemistry. 2002;**12**:1557-1562

[54] Wu M, Chen D, Huang T. Synthesis of au/Pd bimetallic nanoparticles in reverse micelles. Langmuir. 2001;**17**:3877-3883

[55] Simoes M, Baranton S, Coutanceau C. Electrooxidation of sodium borohydride at Pd, au, and Pd_xAu_{1-x} carbon-supported nanocatalysts. Journal of Physical Chemistry C. 2009;**113**:13369-13376

Chapter 5

Synthesis of NPs by Microemulsion Method

Antonio Cid

Abstract

Microemulsions are self-aggregated colloidal systems that provide a controllable system with a promising application as nanoreactors: they can act as pools within which the properties of the nanoparticles can be controlled without difficulty. So in this chapter, I will deal with the metal NPs synthesized by the microemulsion method. This method allows in some cases to control the properties of size, shape, and crystal structure of the metallic NPs, thus generating with the same reagents a series of seeds of different shapes and sizes. The control of the reaction time, the temperature, and the reaction conditions will give us a production of different geometries that will find different applications in large range of research fields.

Keywords: microemulsion method, nanoparticles, colloidal suspensions, self-assembly, surfactant

1. Introduction

During the last decades, self-assembly of micro- and nano-materials attracted a widespread mindfulness because of special properties when comparing with bulk counterparts, because of mechanical and physicochemical property changes at micro- and nanoscale. I will focus on a particular three-phase diagram proportion that is known as the microemulsion system. It is thermodynamically stable and isotropic dispersion of two immiscible liquids, where a surfactant stabilizes micro-domains [1]. It includes different structures and configurations corresponding to different performances of self-assembled colloidal systems. Then I will focus on the assembled structure of nanoparticles. To be able to obtain these structures by microemulsion method, it must be known making a good use of the properties of the microemulsion systems, in which the composition of these structures will determine properties as important as the size of the microdroplets. That ultimately will determine the reaction system of our synthesis of nanoparticles.

The microdroplet environment offers a controllable medium to obtain NPs with tailored size, shape, and structural properties.

2. Metal nanoparticles (MNPs)

2.1 Anionic surfactants

Here I focus on MNPs synthesized by microemulsion method using the anionic surfactant, reviewing literature that employed this surfactant type I found a very noticeable research.

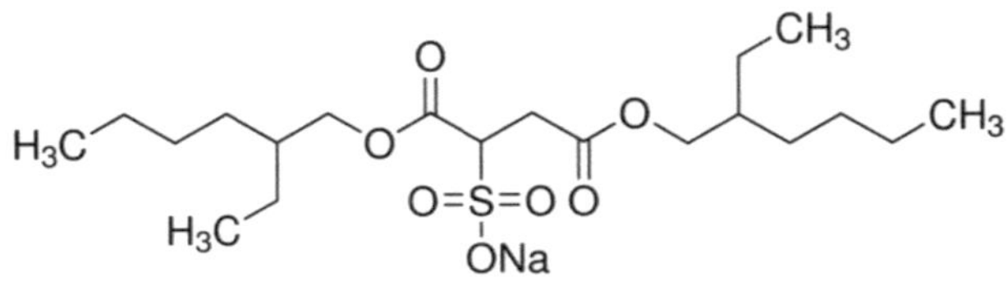

Figure 1.
AOT chemical structure.

Microemulsion method was used by Kon-No et al. [2]; they prepared two microemulsions, one them solubilizing $FeCl_3$ aqueous solution in AOT/cyclohexane and the other aqueous NH_3 in AOT/cyclohexane; both were mixed joint (see AOT chemical structure in **Figure 1**). To obtain the formation of Fe_3O_4 in the micelle, authors added a $FeCl_2$ solution to the mixture of two microemulsions. The solution showed stability during 2 years. A diluted solution of the colloid obtained was characterized by TEM to determine the crystal structure and size distribution by electron micrography. Then the authors obtained magnetic measurements together with an extended analysis to determine detailed surface magnetic properties that allowed them to conclude that (i) distribution function of colloidal particles is a log-normal type, coincidentally with results obtained by gas evaporation technique, (ii) the magnetic interaction between the colloidal particles is negligible, (iii) the colloidal system showed superparamagnetism about 50K, (iv) paramagnetic Fe ions potentially are present on a particle surface, and (v) the temperature dependence of the spontaneous magnetization linearly decreases with the increase of temperature.

López-Quintela and Rivas [3] reported a simple and powerful method to obtain ultrafine particles by means of chemical reactions in microemulsions. The reaction happened within the nanodroplets of the microemulsion. Varying droplets' radii authors were able to control particle dimension. They reported that the proper selection of microemulsion components would ease the exchange of reagents between the "transition dimers" formed by two droplets of the microemulsions; for this to happen, an important alteration in the local curvature is necessary, so the selection of adequate surfactants that shows a radius of curvature close to their natural radius will facilitate the opening of channels [4–9], particle formation, and growth. Once the particles are formed, the surfactant particles act as surface agents, limiting the future growth of the particles. The sizes of the microemulsion droplets can be tuned between 5 and 50 nm by changing the relation of the components of the microemulsion (e.g., $W = [H_2O]/[AOT]$) or varying the microemulsion itself. The results for Fe particles obtained in an AOT microemulsion system ($W = 22.2$; $[AOT] = 0.05$ M; reactant A: $[FeCl_2] = 1.9 \times 10^{-4}$ M; reactant B: $[NaBH_4] = 8.8 \times 10^{-4}$ M) were discriminated using XRD. Nucleation process was confirmed by enhancing the number of scattering centers and hence the scattering intensity. Conversely, growth of the particles is associated with a decrease on the scattering intensity because of the "disappearing" of the smaller particles during their growth. The authors concluded that this general method produces ultrafine particles with tunable size by means of chemical reactions within microemulsions. This simple and reproducible method allowed them to produce quasi-monodisperse particles, within thermodynamically stable system (microemulsion) that was used to control the growth of the particles.

Mann et al. reported [10] that interfacial activity could be used to combine nanoparticle synthesis and self-assembly to produce complex organized materials rising from the interaction of tensioactive molecules attached to specific nanoparticle crystal faces. Authors demonstrated this principle by controlling the $[Ba^{2+}]:[CrO_4^{2-}]$ inside of droplets of the microemulsion that tunes the fusing

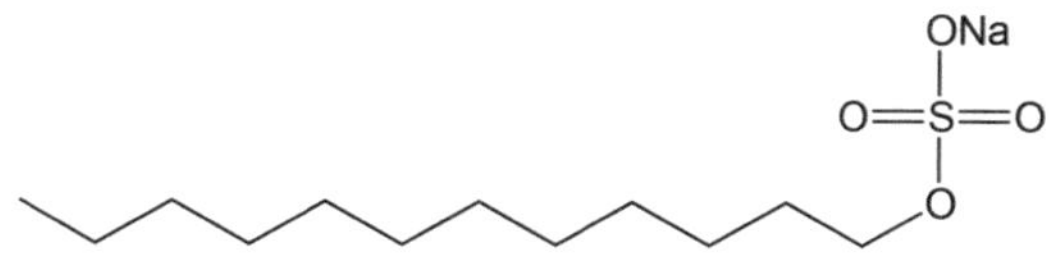

Figure 2.
SDS chemical structure.

of microdroplets and reverse micelles. It allowed them to produce linear chains, rectangular superlattices, and long filaments, as a function of reactant ratio. They carried out the synthesis and self-assembly of NPs by converting sodium AOT to $Ba(AOT)_2$ by direct precipitation [11].

Wang et al. [12] reported by the first time a new microemulsion method to synthesize magnetite NPs. The novelty of their method founded on the cost-effective use of a single microemulsion. The authors prepared the microemulsion by solubilizing NaOH solution into DBS/ethanol/toluene system (DBS, sodium dodecylbenzene sulfonate), and then ferric and ferrous salts with molar ratio (2:1) were made into a mixed solution. A volume of this mixture was drop-wise added to microemulsion under vigorous stirring. The mixed system turned black immediately. After the mixed system was stirred during ½ h and then left undisturbed during 5 h, a two-layer system appeared. The upper layer was black, which contained the NPs, and lower layer was discarded. Following a magnetic separation technique, the magnetite NPs were extracted and then washed several times in ethanol/water solution (1:1) to obtain magnetite NPs.

He et al. reported [13] the chemical synthesis of a nonviral gene transferor, calcium carbonate NP using SDS microemulsions (see SDS chemical structure in **Figure 2**) [14, 15].

2.2 Cationic surfactants

Here I focus on MNPs synthesized by microemulsion method using the cationic surfactant, examining literature that employed this surfactant type I found very outstanding research.

Chin and Yaacob [16] reported the synthesis of magnetic iron oxide nanoparticles by preparing water-in-oil microemulsion system. Authors investigated two different volumetric ratios of Fe^{2+} and OH^- (1:1 and 2:1). Crystallized, physical, and magnetic sizes of magnetic iron oxide nanoparticles were in the superparamagnetic size range (XRD, TEM, and AGM). Superparamagnetic behavior was warranted by hysteresis loop nonappearance in magnetization curve at room temperature. MIONPs were obtained by Massart's method too. The saturation magnetization of MIONPs obtained by *w/o* microemulsion system was larger though the crystallization size was lower, which demonstrated that MIONPs could be custom-made by distinct methods.

The preparation of iron oxide NPs was carried out according to the following route. Ferrous chloride salt solution was prepared (0.2 M); it was steady by adding a few drops of hydrochloric acid (0.5 M). Afterward, the microemulsion was set by melting HTAB in n-octane; subsequently 1-butanol was added followed by $FeCl_2$ aqueous salt solution. The system was gradually shaken till a clear microemulsion suspension was attained. By repeating above method a microemulsion containing 0.25 M NaOH solution was produced.

These microemulsion systems were next fused at 1:1 vol. Instantaneously, a dark green solid was produced before it turned to black. The NPs were gathered and later cleaned repeatedly with acetone and de-ionized water and then they were dried at room temperature which is called S1. The same route was followed to yield a second sample with primary $[Fe^{2+}]$ = 0.1 M. [NaOH] was kept at 0.25 M. The volumetric proportion of microemulsion containing Fe^{2+} to microemulsion containing OH^{-} was 2:1. The sample was called S2 [17]. To compare, by Massart's procedure, S3 has been obtained [18]. Magnetic iron oxide NPs (MIONPs) were effectively formed at r.t. with microemulsion method. The MIONPs showed spherical shape. From the results of TEM, AGM, and XRD, the crystallite, physical, and magnetic sizes of the MIONPs were <10 nm. The mean physical size for S1, S2, and S3 is 6.5, 4.2, and 8.7 nm, respectively. The rise of the total number of microemulsion produced smaller particles. Particles produced by microemulsion procedure were smaller in size and showed higher saturation magnetization, compared to the particles obtained by Massart's procedure.

Shen et al. [19] prepare CTAB-stabilized gold nanowires and nanoparticles with a networked structure and different shapes using in situ *n*-butanol reduction in CTAB/n-butanol/n-heptane/$HAuCl_4$(aq) through microwave dielectric heating (XRD, UV-vis, and TEM). The shape of the hydrophobic gold nanocrystals was effectively modulated tuning of CTAB/$HAuCl_4$ ratios. Anisotropic gold nanostructure formation mechanism was deliberated, which confirmed that CTAB (**Figure 3**) perform a preeminent key role in the drummed-up gold nanowires.

With the decrease of W, the shapes of gold nanocrystals obtained by this method changed from sphere-like decahedron nanoparticles to nanowires with a networked structure. It was revealed that $AuCl^{4-}$ cation and ion of surfactant CTAB played key roles on formation and stabilization of the shape of gold nanowires. The attracting force among gold nanoparticles, which also caused an orienting growth of gold nucleus, was caused by the preferential adsorption of $AuCl^{4-}$ and the selective adsorption of CTAB on the facets of preliminary gold particle surface. This method novel and simple for synthesizing gold nanowires with a networked structure is expected to be appropriate to the synthesis of other metals to attain novel nanostructures.

Sharma et al. [20] reported a detailed and complete study on the growth kinetics of iron oxalate nanorods as they formed throughout several days inside the water pool of CTAB microdroplets (MDs) in isooctane. Authors underlined the novelty of this experimental study on nanostructure preparation in microemulsion-based reactions that could explain the key role of droplet interplay in the reaction kinetics. DLS, TEM, and FCS characterization was completed throughout some days to follow the whole growth kinetics of iron oxalate nanorods beginning from their particle nucleation. Considering the FCS data with an appropriate kinetic model, the droplet-fusion time or dimer lifetime 28 μs was obtained along with the droplet melting rate and the equilibrium constant of the chemical reaction. The droplet association rate exhibited a noteworthy time dependency that straight relates it to the growth mechanism. Combining FCS, DLS, and TEM, three different periods in the complete nanorod growth kinetics appeared: (i) a prolonged nucleation-dominant NP growth period, then (ii) a short period where isotropic NPs shifted to anisotropic growth to form nanorods, and (iii) finally the period where droplet-fusion-assisted elongation of nanorods was proved. The detailed methodologies discussed in this study could be applied to understand the growth kinetics of nanostructures, which are required for various applications in nanoscience and nanotechnology.

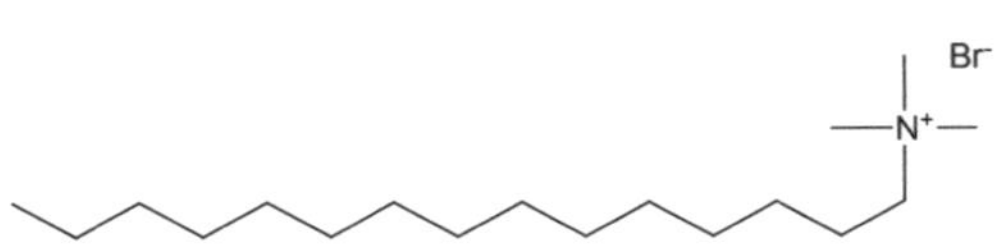

Figure 3.
CTAB chemical structure.

2.3 Nonionic surfactants

Here I review the literature about the use of nonionic surfactants as stabilizer/emulsifier of the microemulsion involved on the synthesis of MNPs by microemulsion method.

Sanchez-Dominguez et al. [21] reported a novel and direct approach to synthesize inorganic NPs at ambient conditions using *o/w* microemulsions to dissolve organometallic precursors. Addition of reducing or oxidizing/precipitating agents implies the formation of metallic or metal oxide nanoparticles, respectively. The authors chose nonionic aggregates and several strategic compositions for nanoparticle synthesis at 25°C. HREM shown that small metal NPs (Pd, Pt, and Rh) and nanocrystalline metal oxide NPs (Ce(IV) oxide with cubic crystalline structure confirmed by XRD) of <7 nm can be obtained at atmospheric pressure and 25°C.

Capek reviewed water-in-oil microemulsion method to prepare metal NPs [22]. This technique allows preparation of ultrafine metal particles ranging in the size diameter between 5 and 50 nm. Author reviewed the previous literature on particle synthesis of various metals such as silver, copper, cadmium, cobalt, nickel, cadmium, and gold in the reverse microemulsion systems. The precursor metal salts and reducing agents are mainly water-soluble molecules, and consequently the MNP nuclei formation progresses in the water microdroplets. The rate of the nuclei formation of the particle depends on the percolation degree of microemulsion droplets. Effects of stabilizer nature and concentration, the oil phase nature, reducing agent, and additive on the NP synthesis are shortened and assessed. The impact of numerous factors such as temperature, the metal salt nature, the incident light, and reaction conditions were also revised.

Tan et al. [23] reported the microemulsion application to prepare silica-coated iron oxide NPs using nonionic surfactants (Triton X-100, Igepal CO-520, and Brij-97, their chemical structure can be observed in **Figures 4–6**, respectively) for the preparation of microemulsions. Iron oxide NPs are formed by co-precipitation.

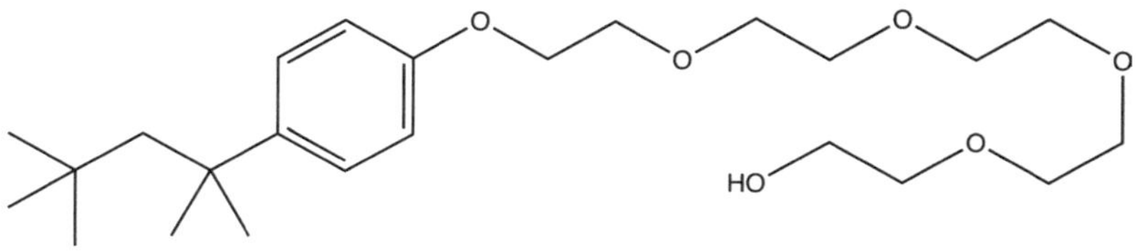

Figure 4.
Triton X-100 chemical structure.

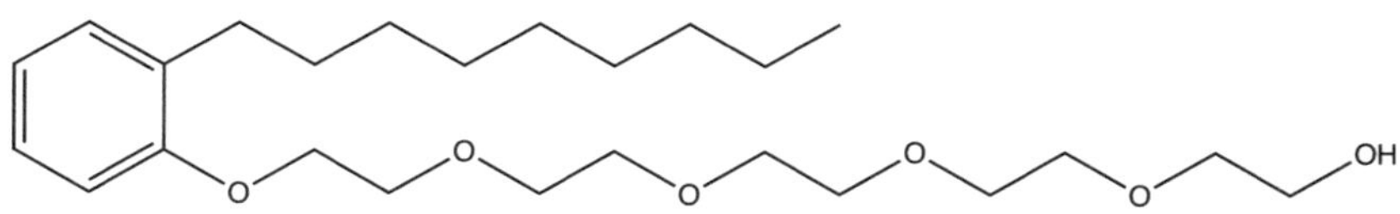

Figure 5.
IGEPAL CO-520 chemical structure.

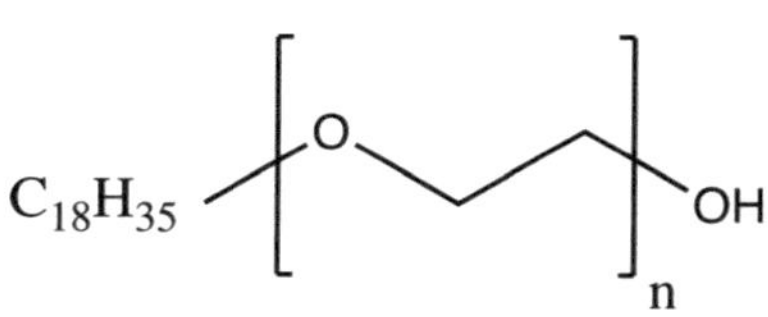

Figure 6.
Brij 97 chemical structure.

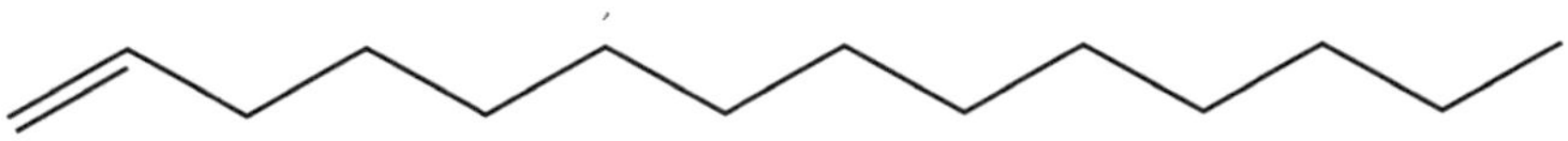

Figure 7.
1-Tetradecane chemical structure.

TEM, XRD, and SQUID magnetometry were used to assess both uncoated and silica-coated FeO NPs. Particles displayed magnetic properties near to that of superparamagnetic materials. By using this route, magnetic NPs ranging 1–2 nm and with very monodispersed in size (percentage std. <10%) were obtained. Base-catalyzed hydrolysis and the polymerization reaction of tetraethyl orthosilicate (TEOS) in microemulsion were used to produce a uniform silica coating of 1 nm enclosing the naked NPs.

Khiev et al. [24] reported the synthesis of regular shape and monodispersed nickel sulfide (NiS) NPs were made in *w/o* microemulsion system including sucrose ester (S-1170) as the nonionic surfactants (which is a nontoxic and biodegradable). NiS NPs were characterized by EFTEM, XPS, and UV-Vis-NIR. NPs showed size ranged between 3 and 12 nm.

The typical microemulsion used in their synthesis had a composition of 28, 56, and 16 wt% of tetradecane/1-butanol (see chemical structure in **Figure 7**) and of aqueous solution containing $Ni(NO_3)_2$ and Na_2S, respectively. The hastened fine particles achieved from centrifugation were cleaned with absolute ethanol and distilled water leastwise fivefold so as to wash away the excess tensioactive byproducts and unreacted reagents. The products were dried in vacuum oven for 16 h at 50°C till a constant weight was attained. It was denoted that the size of the NPs enhanced with the reactants' concentration due to the fusion of nuclei. The presence of quantum confinement effect was apparent for the resultant NPs as the likely bandgap energy exhibited noteworthy increase. Sugar-ester-based nonionic microemulsion offers proper micro-medium to formulate NPs with narrow size distribution and high uniformity.

Pine-Reyes and Olvera reported [25] the use of human and environmentally friendly *w/o* microemulsions to synthesize ZnO NPs based on a not complex procedure (Span 80 and Tween 80 mixture—**Figures 8** and **9**). The method lets achieving NPs with a mean size of approximately 31.2 nm, with low range size variation and pseudo-spherical morphology. XRD, SEM, and TEM confirmed ZnO hexagonal wurtzite phase. The microemulsion organic phase was formed by the mixture of different proportions of emu oil and surfactant (1:1, Span 80, Tween 80) (see their chemical structures in **Figures 8** and **9**, respectively). Next, an aqueous solution of ZnAc (0.5 M) was slowly drop wised to the previous mixture. The systems were mixed under magnetic stirring (1200 rpm, 5 min) and later stored 24 h at r.t. The ZnO powders were synthesized by the drop-wised addition of a (aq) solution of NaOH (1.0 M) to previous prepared microemulsion. NaOH acted as precipitating agent. The mixture was maintained at 60°C and stirred at 1200 rpm for 5 min. In this step, NaOH diffused into the continuous phase achieves the oil-solution interface, which eased the formation of ZnO. Then, mixtures were successively washed by centrifugation at 7500 rpm in water, hexane, and acetone to separate oil and surfactant residues and byproducts from precipitates. Finally, the precipitates were dried during 1 h at 100°C and then calcined by 2 h at 800°C.

Sample M10 with 15% ZnAc, 55% surfactant, and 30% oil exhibits the smallest average particle size, 31.2 nm, and a distribution of 41.2 nm. SEM and TEM micrographs have demonstrated a pseudo-spherical morphology. XRD analysis confirms that samples presented a hexagonal wurtzite phase with preferential growth

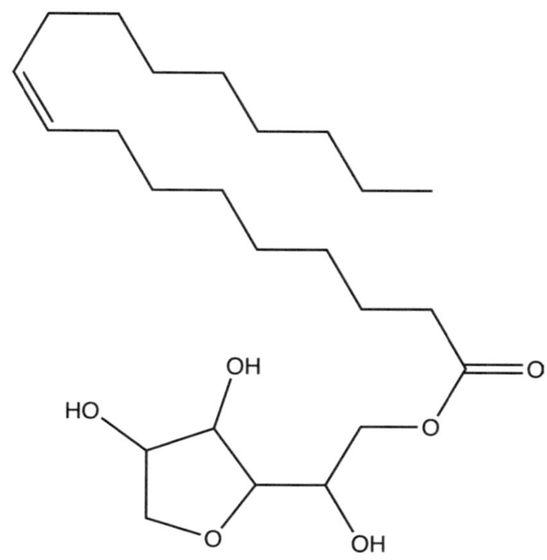

Figure 8.
Span 80 chemical structure.

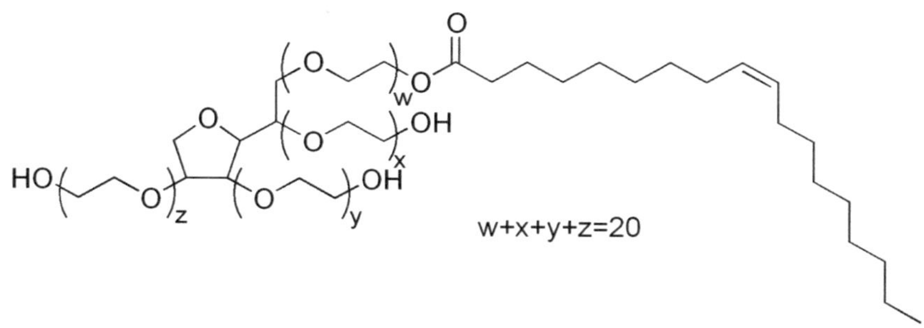

Figure 9.
Tween 80 chemical structure.

direction [101]. The results showed that this method is a suitable pathway for ZnO NPs synthesis. Moreover, it is a simple, direct, and cost-effective method for obtaining nanostructured materials with small particle size and narrow size distribution particle.

3. Silica nanoparticles

Here we focus on a metalloid and the use of microemulsion method to obtain silica NPs. Yamauchi et al. [26] reported the synthesis of ultramicroparticles of silica by means of *w/o* microemulsion method.

Shantz et al. reported [27] synthesis of silicalite-1 nanocrystals in AOT-based microemulsion. They demonstrated that anionic microemulsions drive to essentially distinct crystal patterns than the nonionic [28] or cationic microemulsions formerly explored. The authors concluded that AOT and SDS anions synthesized by microemulsion method were employed to crystallize silicalite-1 nanocrystals with well-defined shapes and structures. Box, disk-like, spherical, and twinned nanocrystals were found under distinct experiential circumstances. It was also shown that tensioactive uniqueness, TPAOH content, and the presence of salts and co-surfactant influence the morphology and crystallinity of silicalite-1. The crystal size and shape contrasted in all cases than those prepared in default of the microemulsion. The crystal morphology could be regulated by adjusting the interplay among surfactant and zeolite surfaces.

4. Outlooks and perspectives

Here I focus on the state-of-the-art research on nanotechnology that involves the microemulsion method.

Chen et al. [29] reported the synthesis of CdS QDs by means of a chemical reaction between amide of cadmium acetate dehydrate and thioacetamide by using a microemulsion-based hydrothermal method. Their properties could be tailored through the use of Emulsifier OP and CTAB, which yield a usual cubic phase and a scarce hexagonal phase, apiece. Authors explored a possible mechanism involving the critical role of surfactant in the formation of crystal structure. A direct dependence of the crystal size that regularly increased with the increase of temperature is also found, and the appearance of red shift in the absorption and emission peaks confirmed the quantum confinement effect. All the desired properties of CdS QDs synthesized by this route denote the chance of the preparation of high-quality QDs under the appropriate reaction conditions.

First, aqueous solutions 0.005 M of cadmium acetate and thioacetamide were set by melting in deionized water with the aid of magnetic stirrer. Next, two reverse microemulsions (M_{Cd} and M_S) with different aqueous phases were prepared. Either M_{Cd} or M_S enclosed three mutual constituents in the volume ratio of 3:5:20, that is, a surfactant of CTAB or emulsifier OP, co-surfactant of n-butyl alcohol, and a continuous oil phase of n-hexane. Once jointly mixed under constant stirring at r.t. during 30 min, it was moved into a 100 mL Teflon-lined autoclave with hydrothermal treatment at 70–120°C for 13 h, succeeded by chilling at r.t. readily. The yellow precipitants were harvested from the synthesis medium by centrifugation, cleaned with anhydrous ethanol and deionized water sometimes and then dried at 70°C, to finally obtain CdS QDs.

The crystal structure was characterized by XRD, obtaining OP at 2θ values of 26.5, 43.9, and 52.1° which agree with the (1 1 1), (2 2 0), and (3 1 1) planes of cubic structure. The morphology was analyzed with a transmission electron microscope, denoting that at 70°C, the mean crystal size of CdS QDs is much tinier (≈3 nm), and afterward it achieved 5.9 nm, when the temperature was increased to 120°C. It noticeably proved the influence of reaction temperature on the crystal size of the as-prepared CdS QDs. The optical properties of samples were characterized by the UV-4100 spectrophotometer and FLS920 steady-state and transient fluorescence spectrometer. In addition, infrared spectroscopy was recorded.

Wei et al. [30] reported the synthesis of the core-shell $NaGdF_4$@$CaCO_3$-PEG NPs were carried out through a facile microemulsion method with using $NaGdF_4$ NPs as templates [31]. Concisely, 5 mL $NaGdF_4$ NPs, 10 mL of cyclohexane, 3.45 mL of Triton X-100, and 3.2 mL of 1-hexanol were added to a flask and combined carefully, followed by annexing 800 μL of a 30 mM solution of $CaCl_2$ to obtain a well-dispersed *w/o* microemulsion. Before the 40 μL, 2.92 M solution of sodium carbonate was added. The microemulsion was kept under mild stirring for 8 h of the mix; the $NaGdF_4$@$CaCO_3$ NPs were gathered by centrifugation and next re-dispersed in 10 mL of ultrapure water. Afterward, 2 mL and 0.2 M PEG8000 (aq.) solution was annexed and stirred for additional 8 h, and subsequently the NPs were gathered by using centrifuge at 10,000 r.p.m. during 15 min and re-dispersed in 20 mL of ultrapure water for further purpose.

In this work, a core-shell nanoparticle of $NaGdF_4@CaCO_3$-PEG was designed as an activable MR/US dual-modal imaging contrast for cancer diagnosis, which is activated by the acidic environment. Authors used the coating of $NaGdF_4$ with a layer of hydrophobic $CaCO_3$ to limit water availability, thus quenching the sphere Gd^{3+} relaxation effects, with the aim of achieving this OFF/ON responsive MR imaging behavior. At acidic aqueous solution, $CaCO_3$ was melted to produce CO_2 bubbles, which is applied to obtain US signal. While a robust MRI augmentation could be triggered on dissolution of $CaCO_3$ and discharge of the earlier quieted $NaGdF_4$ in the aqueous solution. In vivo results confirmed the heavy dual-modal magnetic resonance/ultrasonic imaging capabilities of $NaGdF_4@CaCO_3$-PEG at the tumor area with an acidic environment. Authors expected that their findings might deliver a novel sight for approaches to developing NPs with reactive dual-modal imaging skills. The here-described proof-of-concept nanoparticles with pH-triggered magnetic resonance/ultrasonic dual-modal imaging improvement may assist as a useful guide to advance various molecular imaging strategies for cancer diagnosis in the future.

Majumder and Roy reported the synthesis of mesoporous CeO_2 nanospheres [32] with noticeably raised surface area. It was ready using reversed micelles by a water-in-oil microemulsion method. The structure and semiconducting properties of the NPs are accurately explored using TEM, FESEM, XRD, and UV-Vis. Despite after high-temperature burning, the structural holding of the nanomaterial was evident by EM. Comparing to those of other sensors of the same type, outstanding performances in terms of sensitivity, response-recovery times, and selectivity were found on the distribution of undoped CeO_2 nanospheres for the detection of low-ppm CO. These CO sensors showed 52% sensitivity with a reply time of only 13 s. The sensor parameters were analyzed as a function of both gas concentration and temperature. In addition to that on the scalable and cost-effective synthesis of CeO_2 nanospheres, authors also reported on the fabrication of packaged CO sensors, which could be potentially used for industrial and environmental monitoring purposes. To carry out the material synthesis, all of the chemicals (analytical grade) were applied without further treatment. Primarily, a microemulsion was formed by merging n-hexane, n-butyl alcohol, and diethyl ether in the weight ratio 3:2:1, energetically stirred till the mix turns out to be clear. The quantity of CTAB to reach critical micelle concentration (CMC) was calculated applying the standard conductometric method, where a sequence of microemulsions were examined changing the [CTAB] between 0.2 and 1.5 g. Then, 10 mL of 1 M aqueous solution of $Ce(NO_3)_3$ including 0.65 g of CTAB was annexed to the mix, and then it was agitated again till it showed transparency. Concurrently, another aqueous solution of 10% (w/v) NH_4OH was prepared. The above solutions were merged strongly till it forms a colloidal suspension. The material was gathered by centrifugation at 6000 r.p.m. during 30 min and consecutively washed using deionized water in an ultrasonic bath during 2 h. Lastly, the collected material was dried under vacuum and calcined at 600°C during 4 h.

Wang et al. reported the synthesis and applications of new biomimetic hybrid nanoplatform with theranostic agents [33]. Authors reported the synthesis of this nanoplatform; it was out by the next method: Silica NP core was used to encapsulate fullerene (C60) by a reverse microemulsion method. First, hexanol, Triton X-100, cyclohexane, and deionized water were jointly mixed. A total volume of 2 mL of fullerene in toluene (2 mg mL^{-1}) was then added into the mixture. Subsequently, 60 mL of ammonium hydroxide solution (28 wt%) and 100 mL TEOS were serially annexed, and the sample was rattled at 800 r.p.m. using a mini-stir bar during 24 h at r.t. Finally, the sample was annexed in 30 mL of ethanol to cease the reaction. The obtained sample was centrifuged at 13,800 × g during 10 min to attain fullerene-inserted silica C60S-NPs. After twofold cleaning with deionized water and ethanol, the NPs were suspended in 3 mL of ethanol and followed by the addition 20 mL of

APTMS. The mixture was stirred using a mini-stir bar (200 r.p.m., during 12 h) at room temperature to acquire the APTMS-coated C60S, C60S-A NPs. Lastly, the sample of C60S-A NPs was merged with DPPC liposomes and shacked for 5 h (r.t.) to harvest the phospholipid bilayer-coated C60S, LC60S-NPs. The LC60S NPs were gathered by centrifuging at 13,800 × g for 10 min. The DPPC liposomes were prepared by hydrating a thin DPPC film. Subsequently sonication for 2 min using a Branson 450 Sonifier, the sample of DPPC (1,2-dihexadecanoyl- sn-glycero-3-phosphocholine) liposomes (≈100 nm) was filtered throughout a 0.2 mm filter earlier mixing with the solution of C60S-A nanoparticles. I carried out a research on re-dispersion and self-assembly of C60 pointed special performance and arrangement depending on the solvent nature [34].

Authors concluded that both the encapsulation efficiency (EE) and minimal drug loading content (LC) of theranostic agents are vital calibers of multifunctional NPs for drug transfer. To attain great EE, low drug-to-NP provision ratios have frequently been applied for encapsulation, which outcomes in low drug LC. Authors designed a eukaryotic cell-like nanoplatform or EukaCell, which can be applied to attain great EE (≈100%) and LC (up to ≈87%) of theranostic agents (DOX and ICG). The release of the encapsulated drug can be exactly controlled using NIR laser irradiation to minimize the potential side effects. With the biomimetic eukaryotic cell-like configuration, the EukaCell had an extended half-life and strong stability in blood circulation and could favorably accumulate in tumor following intravenous injection. Finally, the drug-laden EukaCell displayed excellent safety and efficacy for cancer therapy. Authors demonstrated the tremendous potential of their EukaCell for providing theranostic agents to detect and treat cancer.

Tianimoghadam and Salabat reported the formulation of new microemulsion to prepare thiol-functionalized AuNPs [35]. The microemulsion system was obtained by mixing a yellow aqueous solution of 0.03 M of $HAuCl_4$ and a 0.05 M solution of TOAB/toluene. The same microemulsion containing 0.4 M of $NaBH_4$ was prepared independently and drop-wised under energetic stirring into the microemulsion enclosing metal ions. The formation of AuNPs was confirmed by the appearance of a stable light-ruby-red color resulting in microemulsion when the reduced Au(III) ions changed. To form a thiol monolayer surrounding the AuNPs, dodecanethiol (17 mg) was added while kept stirring during 1 h. To eliminate all residual TOAB surfactants (see TOAB chemical structure in **Figure 10**) and thiols from the AuNP surfaces, the AuNPs were pipetted into a 10 mL vial along with 3 mL ethanol and then centrifuged. This procedure was sixfold repeated, after which the dodecanethiol-stabilized AuNPs were then resuspended in toluene. The stable nanocolloid system of Au/toluene was then prepared for characterization. To confirm the size distribution and formation of the AuNPs, TEM techniques and UV-Vis absorption were used. The absorption maximum was found to be 524 nm, which is a shift typical for spherical AuNPs. This surface plasmon band agrees to 3–4 nm particles. According to the Mie theory, the purification procedure removed the

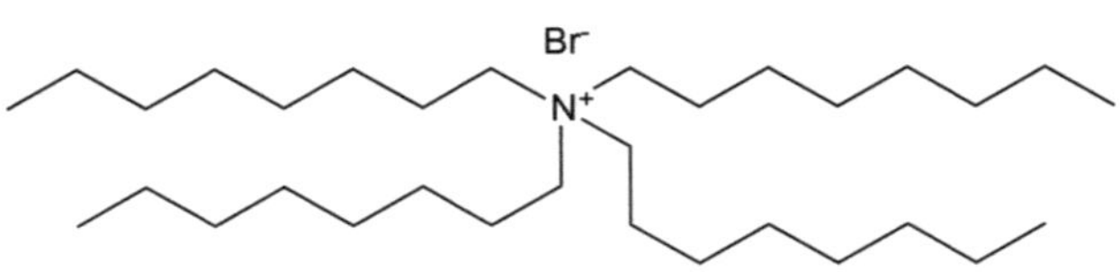

Figure 10.
TOAB chemical structure.

low-molar-mass impurities without essentially varying the structure of the NPs. The AuNPs were basically spherical and monodispersed with diameters around 3–4 nm.

Authors discussed an easy method to prepare dodecanethiol-stabilized AuNPs through a new microemulsion TOAB cationic surfactant-based system. This approach was related with the liquid-liquid phase method. Both AuNPs arranged by different procedures were studied with UV-Vis spectroscopy and TEM imaging, denoting good steadiness. The TEM images of the thiol-capped AuNPs ready by the microemulsion method showed a more acute monodispersity related to that prepared using the Brust method. Since all the methods could be carried out in one step, the microemulsion procedure was faster and easier.

5. Conclusions

As this review chapter has stated, microemulsion method must be taken into account when the aim of the researcher is to prepare well-controlled, narrow-sized, monodispersed NPs.

The influence of microemulsion components has been reviewed; not only the ratio between aqueous phase/surfactant and surfactant/metal precursor but also the influence of the nature of the surfactant—ionic, cationic, and nonionic surfactants—will offer different self-assembled systems. It was underlined that hydrodynamic radius (r_W = 1.5 W) of the microdroplets that forms the microemulsion essentially depends on W = [water/surfactant] so the size of our template will be defined by this relation and microdroplet will offer a well-defined microenvironment to prepare NPs. The influence of the tensioactive nature was also considered by the interaction with metal precursors, co-surfactant, during the nucleation and growth processes and consequent influence upon surficial charge of the NPs. Thus the influence of the reagents and its proportions or ratio will determine the size of water pool within the microdroplets, which finally determine size, shape, morphology, and crystalline structure of the NPs. The control of the reaction time, the temperature, and reagent nature and ratios between them will give us a production of different geometries that will find different applications in wide range of research fields, such as chemical sensors, CO sensing, applications on drug delivery and theranostic, quantum dots, MRI, and biomedicine. It can be concluded that microemulsion method actually offers a good route for synthesis of NPs.

Acknowledgements

Dr. A. Cid acknowledges the postdoctoral contract granted to CIA research group by FEDER Funds at Physical Chemistry Department Universidade de Vigo and the Unidade de Ciências Biomoleculares Aplicadas-UCIBIO which is financed by national funds from FCT/MEC (UID/Multi/04378/2013) and co-financed by the ERDF under the PT2020 Partnership Agreement (POCI-01-0145-FEDER-007728). This chapter is dedicated to Professor Julio Casado Linarejos, who was a great support during my formation as scientist.

Conflict of interest

Author does not declare conflict of interest.

Appendices and nomenclature

NPs	nanoparticles
MNPs	metal nanoparticles
QDs	quantum dots
MIONPs	magnetic iron oxide NPs
AOT	sodium bis(2-ethylhexyl)sulfosuccinate
SDS	sodium dodecyl sulfate
CTAB	cetyltrimethylammonium bromide
HTBA	hexadecyltrimethylammonium bromide
Triton X-100	4-(1,1,3,3-tetramethylbutyl)phenyl-polyethylene glycol
TOAB	tetraoctylammonium bromide
CMC	critical micelle concentration
DBS	sodium dodecylbenzene sulfonate
W/O	water-in-oil microemulsion
O/W	oil-in-water microemulsion
W	water/surfactant ratio
TEM	transmission electron microscopy
AGM	alternating gradient magnetometry
XRD	X-ray diffraction
DLS	dynamic sight scattering
EFTEM	energy filter transmission electron microscopy
XPS	X-ray photoelectron spectroscopy
UV-vis	UV-vis spectroscopy

Author details

Antonio Cid[1,2]*

1 Physical Chemistry Department, Faculty of Sciences, University of Vigo, Ourense, Galicia, Spain

2 UCIBIO@REQUIMTE, Chemistry Department, Faculdade de Ciências e Tecnologia, Universidade Nova de Lisboa, Caparica, Portugal

*Address all correspondence to: acids@fct.unl.pt

References

[1] Lu A, Salabas EL, Schüth F. Magnetic nanoparticles: Synthesis, protection, functionalization, and application. Angewandte Chemie, International Edition. 2007;**46**:1222-1244. DOI: 10.1002/anie.200602866

[2] Bandow S, Kimura K, Kon-No K, Kthara A. Magnetic properties of magnetite ultrafine particles prepared by *W/O* microemulsion method. Japanese Journal of Applied Physics. 1987;**26**(5):713-717

[3] López-Quintela MA, Rivas J. Chemical reactions in microemulsions: A powerfull method to obtain ultrafine particles. Journal of Colloid and Interface Science. 1993;**158**:446-451

[4] Astray G, Cid A, García-Río L, Hervella P, Mejuto JC, Pérez-Lorenzo M. Organic reactivity in AOT-stabilized microemulsions. Progress in Reaction Kinetics and Mechanism. 2008;**33**:81-97. DOI: 10.3184/146867807X273173. ISSN: (1468-6783)

[5] Cid-Samamed A, García-Río L, Fernández D, Mejuto JC, Morales J, Pérez M. Influence of n-alkyl acids on the percolative phenomena in AOT-based microemulsions. Journal of Colloid and Interface Science. 2008;**318**(2):525-529. DOI: 10.1016/j.jcis.2007.11.001. ISSN: (0021-9797)

[6] Serxio IA, Cid A, García-Río L, Mejuto JC, Morales J. Influence of polyethylene glycols on percolative phenomena in AOT microemulsions. Colloid & Polymer Science. 2010;**288**(2):217-221. DOI: 10.1007/s00396-009-2122-0. ISSN: (0303-402X)

[7] Cid A, Astray G, Manso JA, Mejuto JC, Moldes OA. Artificial Intelligence for electrical percolation of AOT-based microemulsions prediction. Tenside, Surfactants, Detergents. 2011;**48**(6):477-483. DOI: 10.3139/113.110155. ISSN: (0932-3414)

[8] Cid A, Gómez-Díaz D, Mejuto JC, Navaza JM. Viscosity and percolative phenomena in AOT based microemulsions. Tenside, Surfactants, Detergents. 2011;**48**(2):165-168. DOI: 10.3139/113.110119. ISSN: (0932-3414)

[9] Montoya IA, Astray G, Cid A, Manso JA, Moldes OA, Mejuto JC. Influence prediction of small organic molecules (ureas and thioureas) upon electrical percolation of AOT-based microemulsions using artificial neural networks. Tenside, Surfactants, Detergents. 2012;**49**(4):316-320. DOI: 10.3139/113.110197. (ISSN: 0932-3414)

[10] Li M, Schnablegger H, Mann S. Coupled synthesis and self-assembly of nanoparticles to give structures with controlled organization. Nature. 1999;**402**:393-395. DOI: 10.1038/46509

[11] Hopwood JD, Mann S. Synthesis of barium sulfate nanoparticles and nanofilaments in reverse micelles and microemulsions. Chemistry of Materials. 1997;**9**:1819-1828

[12] Liu ZL, Wang X, Yao KL, Du GH, Lu QH, Ding ZH, et al. Synthesis of magnetite nanoparticles in *W/O* microemulsion. Journal of Materials Science. 2004;**39**:2633-2636

[13] He H-w, Liu T, Chen YX, Cheng DJ, Li XR, Xiao Y, et al. Calcium carbonate nanoparticle delivering vascular endothelial growth factor-C siRNA effectively inhibits lymphangiogenesis and growth of gastric cancer in vivo. Cancer Gene Therapy. 2008;**15**:193-202. DOI: 10.1038/sj.cgt.7701122

[14] Cid A, Mejuto JC, Orellana PG, López O, Rial R, Simal-Gandara J. Effects of ascorbic acid on the microstructure and properties of

SDS micellar aggregates for potential food applications. Food Research International. 2013;**50**(1):143-148. DOI: 10.1016/j.foodres.2012.10.009

[15] Cid A, Morales J, Mejuto JC, Briz-Cid N, Rial-Otero R, Simal-Gándara J. Thermodynamics of sodium dodecyl sulphate-salicylic acid based micellar systems and their potential use in fruits postharvest. Food Chemistry. 2014;**151**:358-363. DOI: 10.1016/j.foodchem.2013.11.076. (ISSN: 0308-8146, e-ISSN: 1873-7072)

[16] Chin AB, Yaacob II. Synthesis and characterization of magnetic iron oxide nanoparticles via *w/o* microemulsion and Massart's procedure. Journal of Materials Processing Technology. 2007;**191**:235-237

[17] Chin AB, Yaacob II. Synthesis and characterization of iron oxides nanoparticles. Key Engineering Materials. 2006;**306-608**:1115-1120

[18] Bee A, Massart R, Neveu S. Synthesis of very fine maghemite particles. Journal of Magnetism and Magnetic Materials. 1995;**149**:6-9

[19] Shen M, Du Y, Yang P, Jiang L. Morphology control of the fabricated hydrophobic gold nanostructures in *w/o* microemulsion under microwave irradiation. Journal of Physics and Chemistry of Solids. 2005;**66**:1628-1634. DOI: 10.1016/j.jpcs.2005.05.078

[20] Sharma S, Pal N, Chowdhurry PK, Sen S, Ganguli AK. Understanding growth kinetics of nanorods in microemulsion: A combined fluorescence correlation spectroscopy, dynamic light scattering, and electron microscopy study. Journal of the American Chemical Society. 2012;**134**:19677-19684. DOI: 10.1021/ja306556e

[21] Sanchez-Dominguez M, Boutonnet M, Solans C. A novel approach to metal and metal oxide nanoparticle synthesis: The oil-in-water microemulsion reaction method. Journal of Nanoparticle Research. 2009;**11**:1823-1829. DOI: 10.1007/s11051-009-9660-8

[22] Capek I. Preparation of metal nanoparticles in water-in-oil (*w/o*) microemulsions. Advances in Colloid and Interface Science. 2004;**110**:49-74. DOI: 10.1016/j.cis.2004.02.003

[23] Santra S, Tapec R, Theodoropoulou N, Dobson J, Hebard A, Tan W. Synthesis and characterization of silica-coated iron oxide nanoparticles in microemulsion: The effect of nonionic surfactants. Langmuir. 2001;**17**: 2900-2906. DOI: 10.1021/la0008636

[24] khiev PS, Huang NM, Radiman S, Ahmad Md S. Synthesis of NiS nanoparticles using a sugar-ester nonionic water-in-oil microemulsion. Materials Letters. 2004;**58**:762-767. DOI: 10.1016/j.matlet.2003.07.006

[25] Pineda-Reyes AM, Olvera ML. Synthesis of ZnO nanoparticles from water-in-oil (*w/o*) microemulsions. Materials Chemistry and Physics. 2018;**203**:141-147. DOI: 10.1016/j.matchemphys.2017.09.054

[26] Yamauchi H, Ishikawa T, Kondo S. Surface characterization of ultramicro spherical particles of silica prepared by *W/O* microemulsion method. Colloids and Surfaces. 1989;**37**:71-80. DOI: 0166-6622/89

[27] Lee S, Carr CS, Shantz DF. Anionic microemulsion-mediated low temperature synthesis of anisotropic silicalite-1 nanocrystals. Langmuir. 2005;**21**:12031-12036. DOI: 10.1021/la052181u

[28] Astray G, Cid A, Manso JA, Mejuto JC, Moldes OA, Morales J. Basic degradation of 3-keto-carbofuran in the presence of non-ionic self-assembly colloids. Fresenius Environmental Bulletin. 2011;**20**(2):354-357

[29] Chen R, Han B, Yang L, Yang Y, Xu Y, Mai Y. Controllable synthesis and characterization of CdS quantum dots by a microemulsion-mediated hydrothermal method. Journal of Luminescence. 2016;**172**:197-200. DOI: 10.1016/j.jlumin.2015.12.006

[30] Wei Z, Lin X, Wu M, Zhao B, Lin R, Zhang Y, et al. Core-shell $NaGdF_4$@ $CaCO_3$ nanoparticles for enhanced magnetic resonance/ultrasonic dual-modal imaging via tumor acidic micro-enviroment triggering. Scientific Reports. 2017;**7**:5370. DOI: 10.1038/s41598-017-05395-w

[31] Zhou C, Chen T, Wu C, Zhu G, Qiu L, Cui C, et al. Aptamer-$CaCO_3$ nanostructures: A facile, pH-responsive, specific platform for targeted anticancer theranostics. Chemistry, an Asian Journal. 2015;**10**(1):166-171. DOI: 10.1002/asia.201403115

[32] Majumder D, Roy S. Development of low-ppm CO sensors using pristine CeO_2 nanospheres with high surface area. ACS Omega. 2018;**3**:4433-4440. DOI: 10.1021/acsomega.8b00146

[33] Wang H, Agarwal P, Zhao S, Yu J, Lu X, He X. A biomimetic hybrid nanoplatform for encapsulation and precisely controlled delivery of theranostic agents. Nature Communications. 2016;**6**:10081. DOI: 10.1038/ncomms10081

[34] Cid Samamed A, Moldes ÓA, Diniz MS, Rodríguez-González B, Mejuto JC. Redispersion and self-assembly of C 60 fullerene in water and toluene. ACS Omega. 2017;**2**:2368-2373. DOI: 10.1021/acsomega.7b00049. ISSN 2470-1343

[35] Tianimoghadam S, Salabat A. A microemulsion method for preparation of thiol-functionalized gold nanoparticles. Particuology. 2018;**37**: 33-36. DOI: 10.1016/j.partic.2017.05.007

Chapter 6

$CoMo/\gamma\text{-}Al_2O_3$ Catalysts Prepared by Reverse Microemulsion: Synthesis and Characterization

José Luis Munguía-Guillén,
José Antonio de los Reyes-Heredia, Michel Picquart,
Marco Antonio Vera-Ramírez and Tomás Viveros-García

Abstract

A series of $CoMo/\gamma\text{-}Al_2O_3$ catalysts was synthesized by a reverse microemulsion method using 1-butanol as organic agent and cetyltrimethylammonium bromide as surfactant. The aqueous phase was used to form the solution of three corresponding Co, Mo and Al precursor salts. The materials were prepared at different solution concentrations in order to obtain different metal contents. All samples were characterized by X-ray diffraction, Raman spectroscopy, nuclear magnetic resonance and nitrogen physisorption. A chemical species distribution study was performed to establish conditions of preparation and the preponderant species present in solution as a function of pH. The materials obtained present high surface areas which decrease as the metal content (Co + Mo) increases. All samples with the exception of that with the highest metal content were amorphous as shown by X-ray diffraction. By Raman spectroscopy, Mo-O-Mo and MoO_{2t} species were observed in all calcined samples. Mo-O-Co, Al-O-Mo, monomers and heteropolymolybdates were observed for the lower metal content samples, and the formation of $CoMoO_4$ and aluminum molybdate species for the higher metal contents. These results suggest that the materials with lower metal loading have species that are easily sulfidable and provide high activity in hydrodesulfurization reactions. A model for the interaction of the species in the aqueous phase of the micelle is presented.

Keywords: reverse microemulsion, $CoMo/\gamma\text{-}Al_2O_3$ catalysts, chemical species distribution diagrams, catalyst characterization

1. Introduction

Regarding sulfur content in diesel fuels, more stringent environmental regulations have motivated research on new catalysts and novel synthesis methods, to produce highly active hydrodesulfurization (HDS) catalysts [1]. The commercial HDS catalysts are based on MoS_2 promoted by Co or Ni, supported on high-surface area γ-alumina. The most common synthesis procedure involves impregnation of aqueous solutions of Mo and Co (or Ni), followed by drying and calcination steps prior to the activation by a sulfur-containing agent. Generally, it is found that Mo could form a monolayer to prevent Ni or Co species to interact with the support. This helps to avoid the formation of undesired species such as cobalt or nickel aluminates or segregated sulfides [2]. Nevertheless, different preparation sequences or procedures have been investigated since this is not a simple task [3, 4]. Furthermore, Co and Mo form the so-called "CoMoS" phase which has been reported as the active phase in this reaction to remove sulfur from compounds such as dibenzothiophene (DBT) and 4, 6-dimethyl dibenzothiophene (4,6-DMDBT) [4, 5]. This means to synthesize sulfides of 3 slabs stacking (CoMoS type II) [4, 6, 7]. It has also been reported that Co(Ni) MoS species can be obtained, with the use of thiomolybdate precursors in the presence of nonionic surfactants. Moreover, the morphology can be controlled, resulting in an increased activity as compared with more conventional synthesis using ammonium polymolybdates [8]. To carry out the impregnation of the support, solutions of ammonium heptamolybdate tetrahydrate and nickel(II) nitrate hexahydrate are generally used. Firstly, monomeric species such as MoO_4^{2-} are deposited on the carrier. These ions are obtained at pH values between 10 and 12 and MoO_3 species are formed after calcination [9, 10]. After impregnation of molybdenum, Co^{2+} species are impregnated at pH between 2 and 5.96 [11, 12] and CoO_x species are obtained after calcination [12]. Alternative synthesis methods have been tested. For instance, spray pyrolysis allowed the formation of nanosized spherical particles of CoMo sulfides supported on Al_2O_3 that increased the activity of the catalysts, due to weak interaction of CoMo and alumina [2].

A method capable of obtaining nanoparticles with interesting applications as heterogeneous catalysts is inverse microemulsion [13]. The water/oil microemulsion (reverse microemulsion) uses surfactant molecules to stabilize the water/oil interface of nanosized water droplets that are dispersed in an organic solvent. These water droplets consist of an aqueous solution of metal precursors [13, 14]. Recently, reverse microemulsion has been used to prepare NiMo catalysts by precipitation; after calcination and sulfidation, nickel was found decorating the edges of MoS_2. However, after HDS reaction, a significant amount of nickel was segregated, provoking a low activity in comparison with a NiMo/γ-Al_2O_3 catalyst taken as reference [15]. Reverse microemulsion is attractive to extend the actual studies to obtain CoMo/γ-Al_2O_3 catalysts with active species as required for HDS, preparing structured nanoparticles in only one step.

In the microemulsion systems, the interaction between ions in solution and the interface where the surfactant and the organic agent coexist is an important issue. ^{13}C NMR studies have shown that Co (II) is retained at the CTAB-hexanol-water interface, with a 1:1 interaction between Co(II) and hexanol [16]. Also, at low concentrations, octahedral Co(II) complexes are formed. On the other hand, in the preparation of alumina, sodium dodecyl sulfate (SDS, anionic) and cetyltrimethylammonium bromide (CTAB, cationic) have been used as mixtures in different proportions. It was found that the SDS head remains in the alumina network with a decrease in its surface area [17]. Some reports describe the synthesis of unsupported Co(Ni)-Mo-S catalyts for HDS reactions, using surfactants and chelating agents as textural promoters, and the materials obtained show bigger surface areas and a higher catalytic activity than commercial catalysts [8].

The effect of the surfactant on the preparation of CoMo/γ-Al_2O_3 by the microemulsion method was reported [18]. Sodium dodecyl sulfate (SDS) and cetyltrimethylammonium bromide (CTAB) were used and compared. Microemulsions synthesized with SDS provided larger size nanodrops than those obtained with CTAB, with a lower amount of surfactant added. After calcination, the solids prepared with SDS showed the presence of sodium sulfate and had surface areas 50% lower than those obtained with CTAB.

In this work, the preparation of catalysts by a reverse microemulsion method has been undertaken to provide nanostructured-supported metals used in HDS. A series of CoMo/γ-Al_2O_3 catalysts were synthesized and characterized, and the influence of concentration of metals on the calcined solid materials was evaluated. In order to determine the solution conditions to obtain Co and Mo on Al_2O_3, chemical species distribution calculations were made. These calculations allowed us to establish pH and concentrations to employ for the microemulsion preparation and the proper chemical species to be obtained in the solid materials. In a first attempt to correlate the catalyst synthesis parameters with the resultant structural and textural characteristics, characterization of the oxidic phases by atomic absorption, N_2 physisorption, X-ray diffraction (XRD), magic angle spinning-nuclear magnetic resonance (MAS-NMR) and Raman spectroscopy is presented and discussed.

2. Experiment

2.1 Preparation of reverse microemulsions

Five samples were prepared using 1-butanol, water and cetyltrimethylammonium bromide (CTAB), according to the mass percentages shown in **Table 1**. All chemicals were reagent grade from Sigma Aldrich, otherwise it is indicated. **Figure 1** shows a diagram for material synthesis. For each catalyst, the organic phase was 1-butanol (C_4H_9OH), and the aqueous phase was composed of cobalt nitrate ($Co(NO_3)_2$ $6H_2O$), ammonium heptamolybdate (AHM, $(NH_4)_6Mo_7O_{24}$ $4H_2O$) and aluminum nitrate $Al(NO_3)_3$ $9H_2O$ (J. T. Baker). For each microemulsion, the required amount of CTAB ($CH_3(CH_2)_{15}NBr$ $(CH_3)_3$) cationic surfactant was added. During the microemulsion formation, the mixture was continuously stirred with a magnetic stirrer, the CTAB surfactant was added slowly, until it turned from turbid to translucent. The electric conductivity was measured continuously with a LabPro Vernier (model CON-BTA) coupled to a conductivity probe with a sensitivity of ±0.001 S/m. Measurements were conducted at a constant temperature of 20 ± 0.1°C. Critical micelle concentration (cmc) was established for all the samples. This occurred at a point where the conductivity showed a sharp inflection point. Each microemulsion was adjusted to pH = 10 with n-butylamine to generate the required species in solution. Their molar concentrations are shown in the last three columns of **Table 1**. The use of pH 10 in the microemulsions was determined with the development of species distribution diagrams obtained through the Medusa (Make Equilibrium Diagrams Using Sophisticated Algorithms) program that generates distribution curves for species in solution, depicted as the logarithm of the concentration versus pH.

	Microemulsions			Species concentration		
Key	Aqueous phase (%)	1-Butanol (%)	CTAB (%)	Co(II) (mM)	Mo(VI) (mM)	Al(III) (mM)
C1	31.7	47.2	21.1	8.05	20.95	197.61
C2	31.9	47.4	20.7	8.50	22.14	162.43
C3	32.2	47.9	19.9	9.21	23.96	125.52
C4	32.6	48.4	19.0	9.65	25.11	78.95
C5	33.0	49.0	18.0	10.03	26.09	27.34

Table 1.
Nominal compositions for the synthesis of microemulsions.

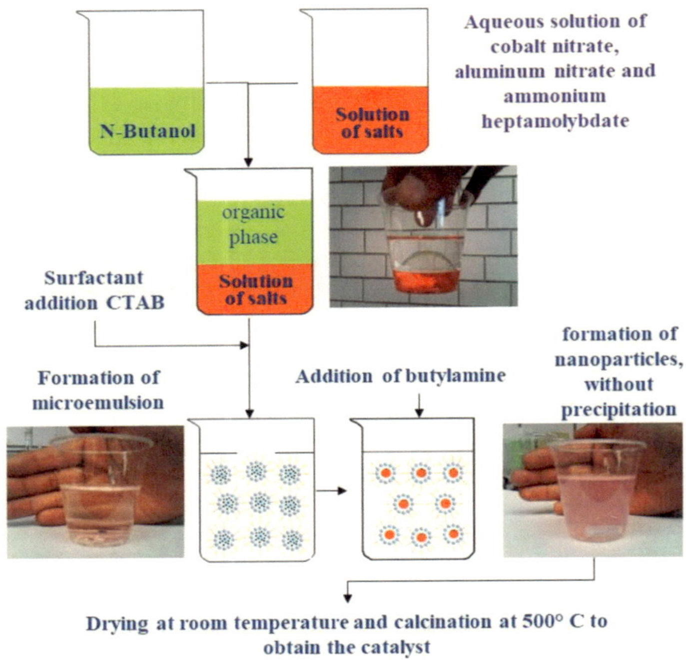

Figure 1.
Diagram for the synthesis method used for preparing the reverse microemulsion.

Subsequent to the formation of the microemulsions, the wet solids were maintained at room temperature for 48 h to evaporate the solvents. The obtained solids were calcined at 500°C for 6 h, with an air flow of 20 ml/min.

2.2 Characterization

The concentrations of the metallic elements in calcined samples were determined in a Varian SpectrAA 220 FS Atomic Absorption Spectrometer equipment. The spectrophotometer was calibrated with certified standards.

The nitrogen physisorption analysis for calcined catalysts was developed in an Autosorb 1 gas sorption system (Quantachrome). Samples were outgassed at 477 K under vacuum for 6 h. Then, nitrogen physisorption experiments were carried out at 77 K. The determination of surface area and volume and pore diameter were carried out using the BET equation and the BJH method, respectively.

To determine the coordination number and crystallographic arrangements of the synthesized support (alumina), a magic angle spinning-nuclear magnetic resonance equipment (MAS-NMR) Bruker Avance II 300, equipped with a multinuclear detector of 4-mm CPMAS with a frequency range ^{31}P to ^{15}N was utilized. Analyses were carried out with a rotational speed of 10 kHz.

The crystalline phases of the calcined samples were identified using a Siemens D-500 Kristalloflex diffractometer, with a CuKα radiation, λ = 0.15406 nm, and with primary and secondary monochromators. The equipment was operated at 35 kV, 20 mA, with a time interval of 1 s and scan rate of 0.03°/s.

Raman spectra for the calcined samples were recorded using Raman HORIBA Jobin Yvon T64000 equipment. The excitation laser source wavelength was 532.1 nm, with a power of 20 mW at the laser head; 100 scans of each sample were performed and accumulated at room temperature, and the spectra were recorded in the range 100–1300 cm^{-1}.

3. Results and discussion

3.1 Determination of species in solution

Figure 2 shows the theoretical diagrams for Co, Mo and Al species as determined from the MEDUSA program for the C3 $CoMo/Al_2O_3$ sample. A total of 28 possible reactions were considered, including 5 soluble compounds (H^+, Co^{2+}, MoO_4^{2-}, Al^{3+}, $C_4H_9NH_2$), 3 solid species ($Al(OH)_3$(cr), $Co(OH)_2$(c), H_2MoO_4(c)), and the other species were Co, Mo or Al complexes.

Figure 2a depicts the distribution of two Co species. One Co^{2+} species appeared at acidic pH and it decreased after a pH value higher than 7. At this value, the formation of $Co(OH)_2$ began to increase, reaching maximum concentration at pH higher than 8.5. Some authors have reported cobalt species in solution pH values higher than 6 [19, 20], in agreement with theoretical calculations. Moreover, these compounds exhibited also high stability.

Seven Mo species were identified in **Figure 2b**. Six of them were observed at acidic pH, among them $HMoO^{4-}$ y $Mo_7O_{24}^{6-}$ species were found at pH values between 4 and 6. MoO_4^{2-} species began to be noticeable at pH values around 4, reaching maximum concentration at pH higher than 6. Furthermore, this was the only remaining species at pH higher than 8. Regarding MoO_4^{2-} species, it has been reported that they exist primarily at low concentrations and at pH = 10 [11]. $[MoO_4]^{2-}$ has also been determined as the predominant oxo-molybdenum (VI) species in the catalyst MoO_3/Al_2O_3, in which case this species is found in solution and adsorbed on the alumina surface depending on the concentration of Mo [21–23].

Regarding Al, **Figure 2c** shows three species. Al^{3+} ions began to disappear at pH values near 2.8 and $Al(OH)_3$ concentration increased. This crystalline species reached its maximum concentration at pH = 4.3 and it decreased at pH = 10.5. The $Al(OH)_4^-$ species began to appear at pH = 11 and its concentration reached a maximum value at pH = 13.7.

Several authors have reported that $Al(OH)_3$ in solution was obtained at pH values between 8 and 11 [17, 24].

Table 2 gives those reactions favoring the most common complexes involved in CoMo catalysts synthesis. These were formed at pH 10. The results from reactions in **Table 2** were considered to establish concentration and pH ranges in which the formation of suitable species within the nanodroplets of the microemulsion may occur. Moreover, these values were taken into account to avoid the release and precipitation of solid particles. This can be observed when an excessive growth of particles within of the micelles takes place. For catalysts with different metal loading (C1, C2 C4 and C5 samples), analogous diagrams were obtained, showing the same species distribution.

Overall, the selected species to be obtained at the conditions fixed experimentally have been those needed for the formation of the precursors of the active species in HDS catalysts [9, 19, 20, 25].

3.2 Atomic absorption

Table 3 gives the results for metallic loadings as determined by atomic absorption for catalysts after calcination. According to **Table 3**, it can be observed that the Co/Mo ratio was around 0.236 for all the samples. Therefore, it was shown that the method of reverse microemulsion allowed us to obtain a Co/Mo ratio with the same average value for all samples, independently of the metal content. Furthermore, the ratio Co/Mo is comparable to metal contents in an industrial catalyst (CoMoind), taken as reference.

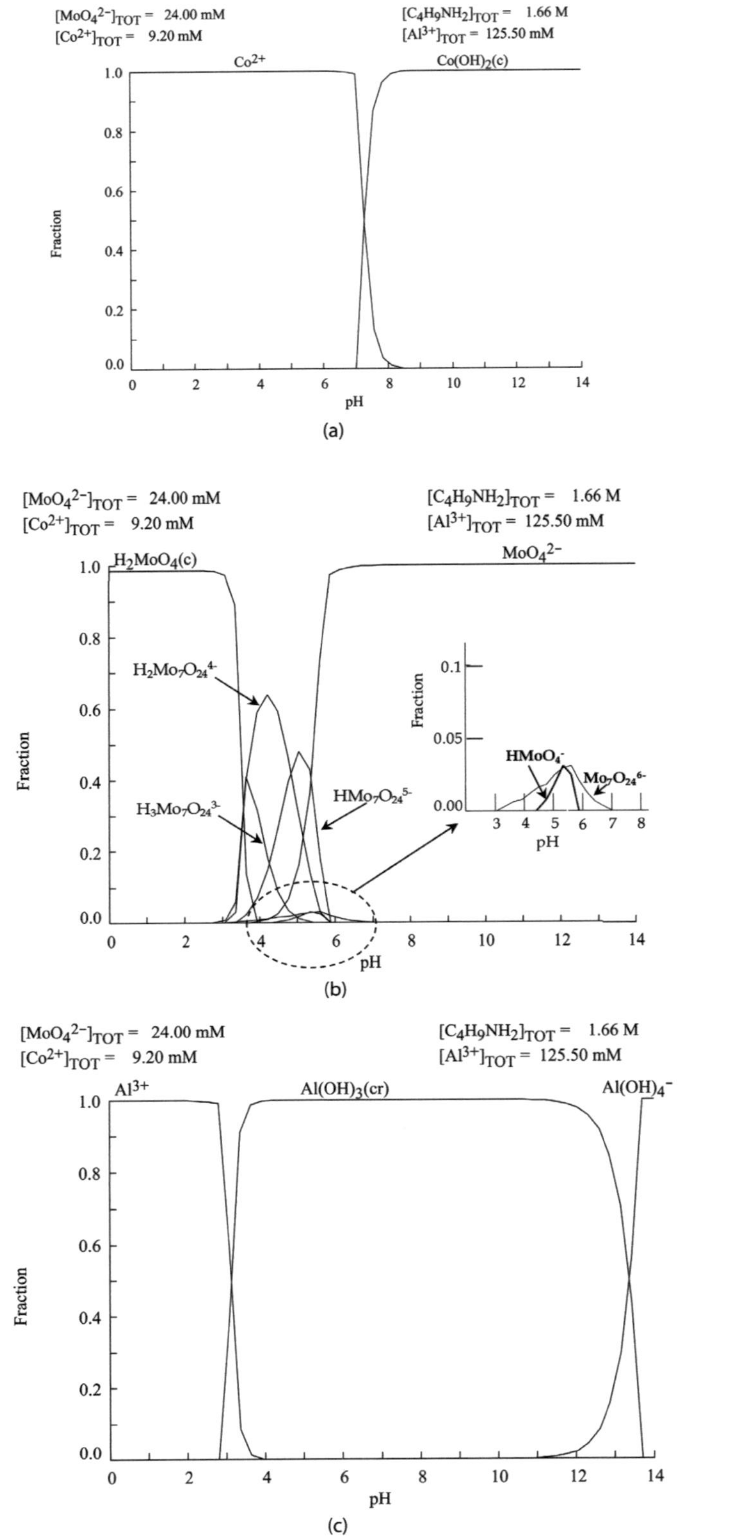

Figure 2.
Species distribution diagrams for the synthesis for CoMo/γ-Al_2O_3 catalysts (C3): (a) cobalt species, (b) molybdenum species (c), aluminum species.

Ions		Formed species	log K
$2H_2O + Co^{2+}$	=	$2H^+ + Co(OH)_{2\,(c)}$	−18.6
$3H_2O + Al^{3+}$	=	$3H^+ + Al(OH)_{3(cr)}$	−8.11
$4H_2O + Mo_7O_{24}^{6-}$	=	$8H^+ + 7MoO_4^{2-}$	−52.99

Table 2.
Hydrolysis reactions for Al, Co and Mo species.

CoMo/γ-Al_2O_3 catalysts have been synthesized by several methods and previous papers [4] agree that the content of metals on the support, including an average ratio of Co/Mo of 0.24, is adequate to prepare active HDS catalysts. However, it has also been reported that metal loading on the support depends on the method of preparation [4]. In some cases, it has been determined that the number of slabs on the support (alumina) increases and the edges on the slabs decrease with increasing Mo content. These edges are generally decorated by deposition of cobalt species as promoter [7]. Therefore, the metal loading must be controlled to an extent, that is, 15 or 20 wt.% of Mo, since high loadings lead to the formation of inactive structures [4]. Furthermore, some studies have been aimed at obtaining a monolayer of molybdenum species on the surface of the carrier and Co/(Co + Mo) ratios corresponding to high dispersion of Co on the edges of the slabs of molybdenum [25, 26]. In our case, the results indicate that precipitation of the particles out of the micelles of the microemulsion systems did not occur during catalyst preparation, since it would have resulted in a heterogeneous distribution of the catalyst composition.

3.3 Nitrogen physisorption

Results for nitrogen physisorption for the catalysts in this work and for an industrial catalyst are given in **Table 4**. Catalysts C1, C2 and C3 exhibited higher surface areas than the industrial catalyst (CoMoind). Catalyst C4 showed a comparable surface area as that of the CoMoind sample, while the C5 catalyst exhibited a significantly lower surface area. Furthermore, pore diameter values increased from catalyst C1 to C5, showing a comparable value between C2 and CoMoind. One of the fundamental aspects of the analysis of physisorption that has been suggested for HDS supports is that their surface area must be high enough to ensure a high dispersion of the active species [2].

Thus, we observed high surface areas for the synthesized catalysts by using microemulsions. It is possible that nanosized particles were formed inside the micelles systems as reported in the literature [27–30]. However, high Mo loading such as those in C4 and C5 catalysts could lead to different porous structures and one cannot rule out pore blocking by the metals.

Key	Co (wt.%)	Mo (wt.%)	Co/Mo
C1	3.7	15.3	0.236
C2	4.6	18.6	0.234
C3	5.6	24.0	0.238
C4	8.0	33.1	0.238
C5	13.2	54.6	0.235
CoMoind	5.7	24.5	0.233

Table 3.
Metal content for the catalysts as determined by atomic absorption.

3.4 Spinning-nuclear magnetic resonance equipment (MAS-NMR)

Figure 3 shows the ^{27}Al MAS-NMR spectrum for the C3 catalyst before calcination. One can observe a broad band with a maximum at around 0 ppm, that can be ascribed to the formation of the $Al(OH)_3$ (gibbsite) compound. This species has low electronegativity and, therefore, it can interact more easily with more electronegative elements [31], such as molybdenum species, rather than with cobalt species. This is relevant for the preparation method, considering that the latter has a tendency to form cobalt aluminate with the support when using impregnation methods [3, 32, 33].

Figure 4 displays the ^{27}Al MAS-NMR spectra for the calcined catalysts. The CoMoind sample exhibited a broad band at 0 ppm that can be assigned to octahedral species of the support. Besides, formation of tetrahedral species was determined as a weak band at around 60 ppm. These two bands are characteristic of γ-Al_2O_3 [34, 35]. It is likely that during calcination up to 250°C, the gibbsite formed boehmite (α-AlO(OH)). This compound was transformed to γ-Al_2O_3, when the temperature reached 500°C [36].

Additionally, **Figure 4** shows that pentahedral species between 40 and 50 ppm appeared for the C5 and CoMoind catalysts. These species have been identified as defects in the support structure, as originated by the replacement of oxygen in the network of octahedral symmetry by hydroxyl groups [37]. For C5 sample, a band around −14 ppm was detected. This band has been related to the presence of the $Al_2(MoO_4)_3$ species which distorts the octahedral network of the support [35].

Some authors have reported that the $Al_2(MoO_4)_3$ species can be due to the dissolution of the Al^{3+} species, which subsequently react with heptamolybdate complexes during the impregnation step, forming the Anderson-type heteropolymolybdate $[Al(OH)_6Mo_6O_{18}]^{3-}$ [35]. In this study, the appearance of the molybdate species (around −15 ppm) could be due to the formation of an Anderson-type heteropolymolybdate obtained in the synthesis mixture.

3.5 X-ray diffraction

Figures 5 and **6** show the diffractograms for the calcined catalysts. As it can be seen, no XRD lines were observed for almost all catalysts, except for the C5 solid, pointing out to amorphous solids. Thus, cobalt and molybdenum oxides may be highly dispersed on the support (γ-Al_2O_3). This has also been observed by other authors who detected a broad band between 5 and 50° (2Θ) describing amorphous catalysts. **Figure 6** (extended from **Figure 5**) shows the diffractogram for the C5 catalyst. Well-defined XRD lines were detected and they were associated with $CoMoO_4$ and $Al_2(MoO_4)_3$ [2]. This finding is consistent with MAS-RMN results for the highly loaded CoMo catalyst (C5).

Catalysts Tc (500°C)	S_{BET} (m^2/g)	Pore volume (cm^3/g)	D_p (Å)
C1	294	0.74	55.8
C2	286	0.62	56.2
C3	261	0.41	68.3
C4	212	0.28	71.2
C5	125	0.19	76.2
CoMoind	204	0.49	56.6

Table 4.
Nitrogen physisorption results for the calcined catalysts.

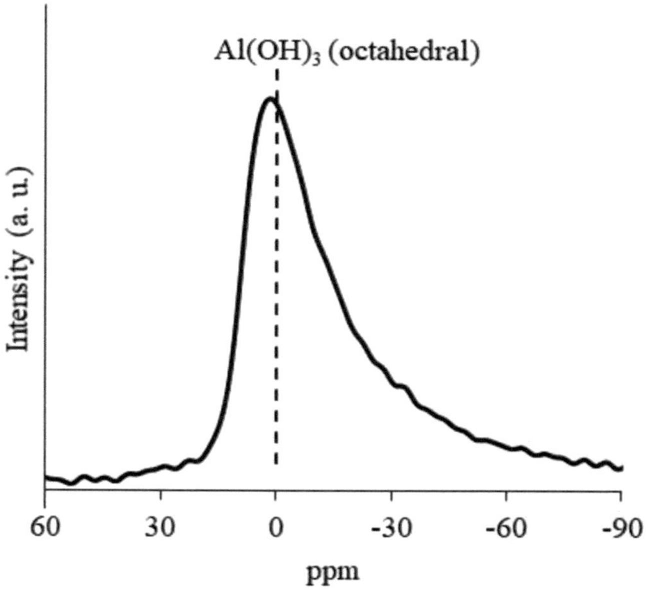

Figure 3.
^{27}Al MAS-NMR spectrum for the C3 CoMo/γ-Al_2O_3 uncalcined catalyst.

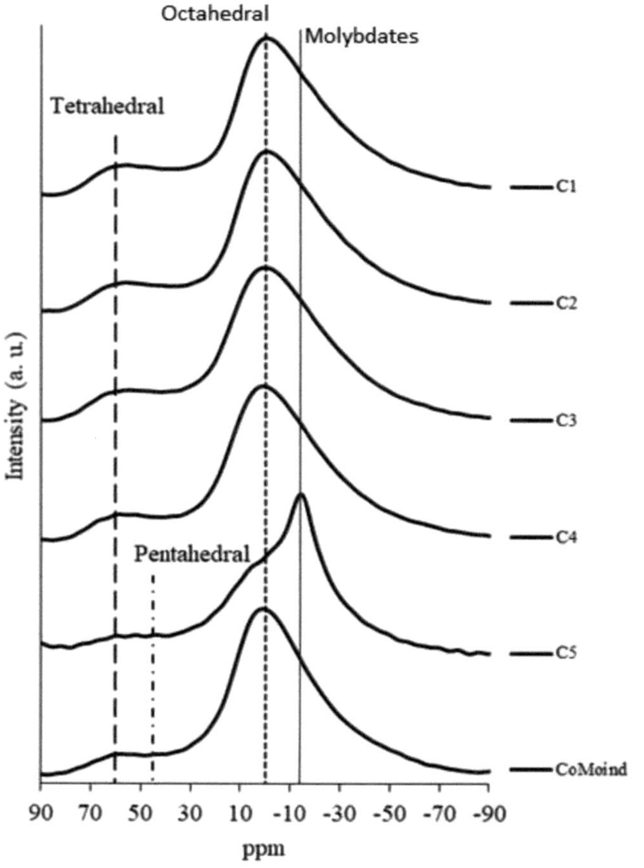

Figure 4.
^{27}Al MAS-NMR spectra for the CoMo/γ-Al_2O_3 calcined catalysts series.

3.6 Raman analysis

Raman spectra for the CoMo/γ-Al_2O_3 catalysts series and the CoMoind catalyst are given in **Figure 7**. For C1, C2 and C3 CoMo catalysts, more attenuated and wider bands were observed, as compared with C4 and C5 spectra. It could be due to more microcrystalline particles in low-content Mo samples. Thus, particle sizes for highly loaded catalysts C4 and C5 were larger, as expected.

Table 5 summarizes all the Raman bands indicating the type of species as assigned for the catalysts in oxidic state. To assign the species present in the samples, the signals were deconvoluted and identified according to those reported in the literature.

One can notice bands between 500 and 700 cm^{-1} for C1 and C2 catalysts. These peaks can be assigned to the stretching mode vibration for bridged Mo-O-M links [38]. Specifically, the presence of Mo-O-Co bonds between 540 and 560 cm^{-1} was identified. This type of band corresponds to the interval of heteropolymolybdate structures and they indicate a strong interaction between cobalt and molybdenum oxides. This means that a weak interaction with the support occurs and, thus, it can induce a high degree of sulfidation of the catalyst as published by others [2, 39]. Besides, bands at 817–818 cm^{-1} were observed for samples with Mo loadings >20% in catalysts C4 and C5. These peaks are generated by MoO_3 species, indicating that a monolayer of MoO_3 on the surface of the support has been exceeded [40, 41]. Moreover, these MoOx species have been identified as orthorhombic molybdate species [42, 41]. Bands between 850 and 875 cm^{-1} were attributed to Mo-O-Mo bonds, assigned to the asymmetric vibrational stretching mode [2, 41, 43]. Other bands located between 930 and 960 cm^{-1} were assigned to Mo=O links, as the vibrational stretching mode for the dioxo groups in oxomolybdate species. This indicates the formation of MoO_{2t} species, where t indicates terminal oxygen atoms. This type of species was present in all studied catalysts, as tetrahedral MoOx structures on alumina [40].

Furthermore, it was found that the bands corresponding to the Mo-O-Mo links increased proportionally when increasing the bands corresponding to the MoO_{2t} links [2, 41, 43]. This suggests that there was an increase in the number of bridges of Mo, which implied high dispersion of Mo on the support [40, 44].

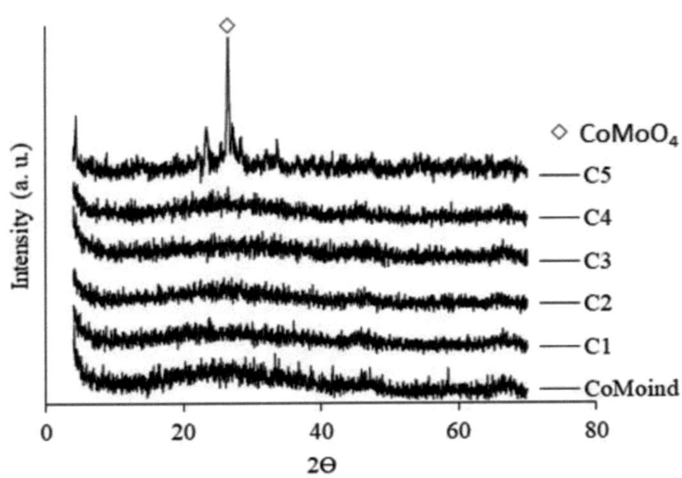

Figure 5.
XRD results for the CoMo/γ-Al_2O_3 calcined catalysts series.

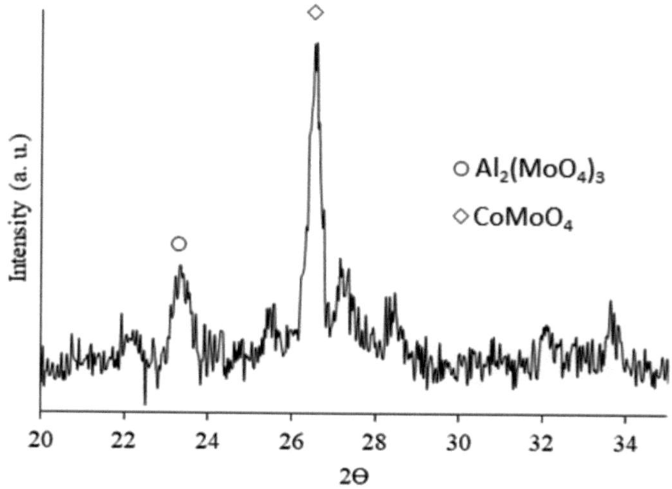

Figure 6.
XRD results for the CoMo/γ-Al_2O_3 calcined C5 catalyst.

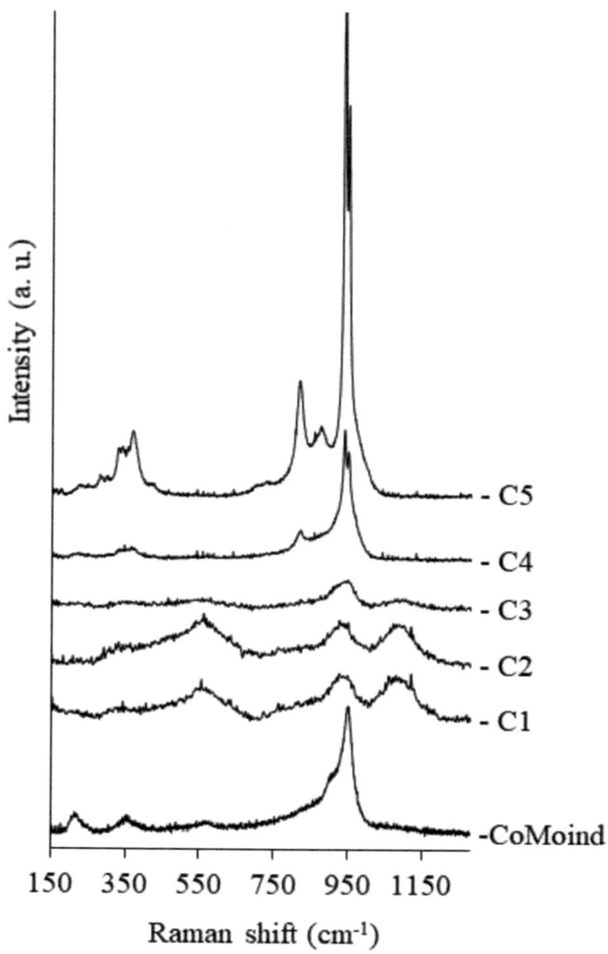

Figure 7.
Raman spectra for the CoMo/γ-Al_2O_3 calcined catalysts series.

Range (cm^{-1})	C1	C2	C3	C4	C5	CoMoind	Assigned species
200–300						215	Heptamolybdates or octamolybdates
300–400	320	320	360	335–365	346–369	354	Monomers Heptamolybdates or octamolybdates
500–700	570	570				560	Mo-O-Co Al-O-Mo
800–900	850	850	870	818 870	817 875	850	MoO_3 Mo-O-Mo
900–1000	937	930	930, 950	939, 950 978	905 938, 949 978	918 951	$CoMoO_4$ $^*MoO_{2t}$ $Al_2(MoO_4)_3$
1000–1100	1085	1086	1090			1045	$(MoO_4)^{2-}$

Table 5.
Raman bands and the species associated for the calcined CoMo/γ-Al_2O_3 catalysts.

Additionally, bands between 905 and 918 cm^{-1} for the C5 and CoMoind catalysts were attributed to the presence of molybdate $CoMoO_4$ species, either isolated or polymerized, involving a strong interaction with the support. As reported by some authors [33, 42, 45, 46], the formation of aluminum molybdate species occurs at high Mo loadings. It has been published also that some amount of Mo reacts with Co during calcination.

For C1 and C2 catalysts, a band in the range between 970 and 1100 cm^{-1} was detected and highly dispersed MoO_4 tetrahedral species can be assigned in agreement with MAS-NMR results. It is likely that these metal loadings did not reach the MoOx monolayer formation [44, 47]. Moreover, the bands appearing between 200 and 400 cm^{-1} were attributed to Mo monomeric species [33].

Regarding Co species, no bands associated with this oxide were identified. However, since no segregation of Co species was noticed, one can propose that there exists an interaction of Mo and Co.

3.7 A model for the interaction of species in solution and the micelle

The results obtained in this work for the characterization of the catalytic materials demonstrate that these have: high surface areas, amorphous γ-Al_2O_3, a MoO_3 type species on the surface of the support and a constant Co/Mo ratio independent of the metal loading. Overall, it is possible to establish an interaction between Co species at the interface with the micelles, preventing the migration of Co into the alumina network. There is also an interaction between Mo and AlOOH in solution which hindered the Co-Al interaction and promoted the formation fo Mo-O-Co, as shown by Raman results. It is possible to depict a representative scheme of the species inside the micelles as can be observed in **Figure 8**. In this diagram, 1-butanol replaces water in the first coordination sphere of Co ion, so that Co is retained at the interface of the micelles [16].

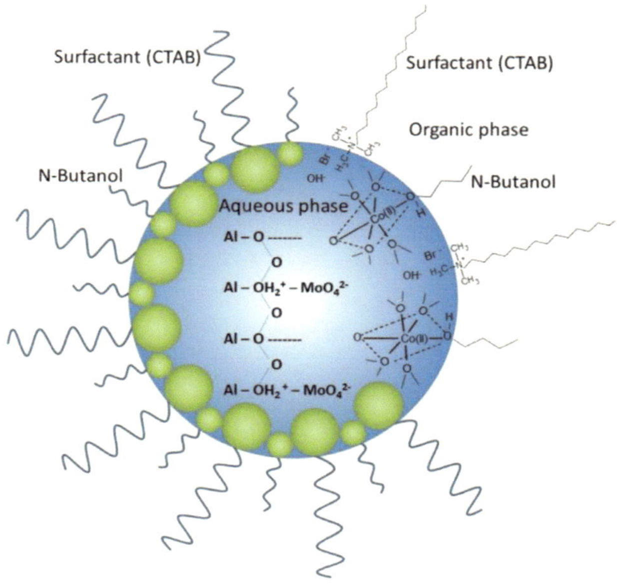

Figure 8.
Schematic diagram for the interaction between species in solution within the micelle. Al species interacting with Mo species, Co species located at the interface.

4. Conclusions

CoMo/γ-Al_2O_3 catalysts were synthesized at several metal (Co + Mo) contents employing a reverse microemulsion method. The microemulsion was formed using 1-butanol as organic agent, cetyltrimethylammonium bromide as surfactant and water. A study of chemical species distribution in solution as a function of pH was performed and provided the pH and precursor salt concentrations to be used in the synthesis to obtain the desired final material. The study allowed to obtain stable micelles, no precipitation of the metallic particles outside the micelles occurred and a constant Co/Mo ratio in all samples independent of the metal loading was observed. The solids calcined at 500°C showed large surface areas which decrease as the metal content was increased. All the calcined samples were amorphous for X-ray diffraction and only at the highest Co + Mo concentration some crystalline phases were found. On these samples, species such as Mo-O-Mo, MoO_{2t}, Mo-O-Co, Al-O-Mo were detected. These species are considered precursors of highly active catalytic sites in HDS reactions. Based on these results, a schematic model for the micelle formed was produced. In this model, Al and Mo species in solution interact whereas Co species interact with 1-butanol at the interface. With this model, it is possible to envisage the formation of the solid material with Mo covering the surface of Al_2O_3 and Co interacting with Mo on the surface of the aluminum oxide.

Acknowledgements

The authors would like to thank CONACYT-Mexico and UAM for the financial support. J. L. Munguía thanks CONACYT-Mexico for the scholarship granted.

Author details

José Luis Munguía-Guillén[1], José Antonio de los Reyes-Heredia[1], Michel Picquart[2], Marco Antonio Vera-Ramírez[3] and Tomás Viveros-García[1]*

1 Department of Process and Hydraulics Engineering, Basic Sciences and Engineering Division, Universidad Autónoma Metropolitana, Iztapalapa, Ciudad de México, México

2 Department of Physics, Basic Sciences and Engineering Division, Universidad Autónoma Metropolitana, Iztapalapa, Ciudad de México, México

3 Department of Chemistry, Basic Sciences and Engineering Division, Universidad Autónoma Metropolitana, Iztapalapa, Ciudad de México, México

*Address all correspondence to: tvig@xanum.uam.mx

References

[1] Babich IV, Moulijn JA. Science and technology of novel processes for deep desulfurization of oil refinery streams: A review. Fuel. 2003;**82**:607-631. DOI: 10.1016/s0016-2361(02)00324-1

[2] Choi K-H, Korai Y, Mochida I. Preparation and characterization of nano-sized $CoMo/Al_2O_3$ catalyst for hydrodesulfurization. Applied Catalysis A: General. 2004;**260**:229-236. DOI: 10.1016/j.apcata.2003.10.019

[3] Papadopoulou C, Vakros J, Matralis HK, Kordulis C, Lycourghiotis A. On the relationship between the preparation method and the physicochemical and catalitic properties of the $CoMo/Al_2O_3$ hydrodesulfurization catalysts. Journal of Colloid and Interface Science. 2003;**261**:146-153. DOI: 10.1016/s0021-9797(02)00167-4

[4] Topsoe H, Clausen BS, Massoth FE. Hydrotreating Catalysis, 1st ed. Berlin Heidelbergh: Springer-Verlag; 1996. 310 p. DOl: 10.1007/978-3-642-61040-0_1

[5] Choi KH, Kunisada N, Korai Y, Mochida I, Nakano K. Facile ultra-deep desulfurization of gas oil through two-stage or -layer catalyst bed. Catalysis Today. 2003;**86**:277-286. DOI: 10.1016/S0920-5861(03)00413-9

[6] Carlsson A, Brorson M, Topsoe H. Morphology of WS_2 nanoclusters in WS_2/C hydrodesulfurization catalysts revealed by high-angle annular dark-field scanning transmission electron microscopy (HAADF-STEM) imaging. Journal of Catalysis. 2004;**227**:530-536. DOI: 10.1016/j.jcat.2004.08.031

[7] Lauritsen JV, Nyberg M, Norskov JK, Clausen BS, Topsoe H, Laegsgaard E, et al. Hydrodesulfurization reaction pathways on MoS_2 nanoclusters revealed by scanning tunneling microscopy. Journal of Catalysis. 2004;**224**:94-106. DOI: 10.1016/j.jcat.2004.02.009

[8] Genuit D, Afanasiev P, Vrinat M. Solution syntheses of unsupported Co(Ni)–Mo–S hydrotreating catalysts. Journal of Catalysis. 2005;**235**:302-317. DOI: 10.1016/j.jcat.2005.08.016

[9] Spanos N, Vordonis L, Kordulis C, Lycourghiotis A. Molybdenum-oxo species deposited on alumina by adsorption: I. Mechanism of the adsorption. Journal of Catalysis. 1990;**124**:301-314. DOI: 10.1016/0021-9517(90)90179-n

[10] Wang L, Hall WK. The preparation and genesis of molybdena-alumina and related catalyst systems. Journal of Catalysis. 1982;**77**:232-241. DOI: 10.1016/0021-9517(82)90163-4

[11] Bergwerff JA, Visser T, Weckhuysen BM. On the interaction between Co- and Mo-complexes in impregnation solutions used for the preparation of Al_2O_3-supported HDS catalysts: A combined Raman/UV-vis-NIR spectroscopy study. Catalysis Today. 2008;**130**:117-125. DOI: 10.1016/j.cattod.2007.06.037

[12] Vakros J, Papadopoulou C, Voyiatzis GA, Lycourghiotis A, Kordulis C. Modification of the preparation procedure for increasing the hydrodesulfurisation activity of the CoMo/γ-Al_2O_3 catalysts. Catalysis Today. 2007;**127**:85-91. DOI: 10.1016/j.cattod.2007.02.028

[13] Eriksson S, Nylén U, Rojas S, Boutonnet M. Preparation of catalysts from microemulsions and their applications in heterogeneous catalysis. Applied Catalysis A: General. 2004;**265**:207-219. DOI: 10.1016/j.apcata.2004.01.014

[14] Uskokovic V, Drofenik M. Synthesis of materials within reverse micelles. Surface Review and Letters.

2005;**12**:239-277. DOI: 10.1142/S0218625X05007001

[15] Scott CE, Pérez-Zurita MJ, Carbognani LA, Molero H, Vitale G, Guzman HJ, et al. Preparation of NiMoS nanoparticles for hydrotreating. Catalysis Today. 2015;**250**:21-27. DOI: 10.1016/j.cattod.2014.07.033

[16] Nagy JB. Preparation of ultrafine particles of metals and metal borides in microemulsions. In: Kumar P, Mittal KL, editors. Handbook of Microemulsion Science and Technology. 1st ed. New York: Marcel Dekker AG; 1999. pp. 499-547. ISBN: 0-8247-1979-4

[17] Sicard L, Lebeau B, Patarin J, Kolenda F. Synthesis of mesostructured or mesoporous aluminas in the presence of surfactants. Comprehension of the mechanisms of formation. Oil & Gas Science and Technology. 2003;**58**:557-569. DOI: 10.2516/ogst:2003039

[18] Munguía-Guillén JL, Vernon-Carter EJ, De los Reyes-Heredia JA, Viveros-García T. Effect of surfactant in the synthesis of CoMo/Al_2O_3 catalysts obtained by reverse microemulsion for dibenzothiophene hydrodesulfurization. Revista Mexicana de Ingeniería Química. 2016;**15**:893-902. ISSN electronic: 2395-8472

[19] Vakros J, Bourikas K, Kordulis C, Lycourghiotis A. Structure of Co(II) species formed on the surface of γ-alumina upon interfacial deposition. The Open Catalysis Journal. 2014;7:8-17. DOI: 10.2174/1876214X01407010008

[20] Towle SN, Bargar JR, Brown GE, Parks GA. Surface precipitation of Co (II)(aq) on Al_2O_3. Journal of Colloid and Interface Science. 1997;**187**:62-82. DOI: 10.1006/jcis.1996.4539

[21] Blanchard P, Mauchausse C, Payen E, Grimblot J, Poulet O, Boisdron N, et al. Preparation and characterization of CoMo/Al_2O_3 HDS catalysts: Effects of a complexing agent. Studies in Surface Science and Catalysis. 1995;**91**:1037-1049. DOI: 10.1016/s0167-2991(06)81847-1

[22] Blanchard P, Lamonier C, Griboval A, Payen E. New insight in the preparation of alumina supported hydrotreatment oxidic precursors: A molecular approach. Applied Catalysis, A: General. 2007;**322**:33-45. DOI: 10.1016/j.apcata.2007.01.018

[23] Iannibello A, Mitchell PCH. Preparative chemistry of cobalt-molibdenum/alumina catalysts. Studies in Surface Science and Catalysis. 1979;**3**:469-478. DOI: 10.1016/s0167-2991(09)60230-5

[24] Parida KM, Pradhan AC, Das J, Sahu N. Synthesis and characterization of nano-sized porous gamma-alumina by control precipitation method. Materials Chemistry and Physics. 2009;**113**:244-248. DOI: 10.1016/j.matchemphys.2008.07.076

[25] Bezverkhyy I, Afanasiev P, Lacroix M. Promotion of highly loaded MoS_2/Al_2O_3 hydrodesulfurization catalysts prepared in aqueous solution. Journal of Catalysis. 2005;**230**:133-139. DOI: 10.1016/j.jcat.2004.12.009

[26] Pérez De la Rosa M, Texier S, Berhault G, Camacho A, Yacamán MJ, Mehta A, et al. Structural studies of catalytically satabilized model and industrial-supported hydrodesulfurization catalysts. Journal of Catalysis. 2004;**225**:288-299. DOI: 10.1016/s0021-9517(04)00161-7

[27] Corolleur C, Tomanova D, Gault FG. The mechanisms of hydrogenolysis and isomerization of hydrocarbons on metals: VII. Isomerization of labeled hexanes and hydrogenolysis of methyl (^{13}C) cyclopentane on a 10% platinum-alumina catalyst. Journal of Catalysis. 1972;**24**:401-416. DOI: 10.1016/0021-9517(72)90124-8

[28] Boutonnet M, Kizling J, Mintsa-Eya V, Choplin A, Touroude R, Maire G, et al. Monodisperse colloidal metal particles from nonaqueous solutions: Catalytic behavior in hydrogenation of but-1-ene of platinum, palladium, and rhodium particles supported on pumice. Journal of Catalysis. 1987;**103**:95-104. DOI: 10.1016/0021-9517(87)90096-0

[29] Dutta P, Fendler JH. Preparation of cadmium sulfide nanoparticles in self-reproducing reversed micelles. Journal of Colloid and Interface Science. 2002;**247**:47-53. DOI: 10.1006/jcis.2001.8097

[30] Hayashi H, Chen LZ, Tago T, Kishida M, Wakabayashi K. Catalytic properties of Fe/SiO_2 catalysts prepared using microemulsion for CO hydrogenation. Applied Catalysis A: General. 2002;**231**:81-89. DOI: 10.1016/s0926-860x(01)00948-6

[31] Hu Y, Liu X, Xu Z. Role of cristal structure in flotation separation of diaspore from kaolinite, pyrophyllite and illite. Minerals Engineering. 2003;**16**:219-227. DOI: 10.1016/s0892-6875(02)00368-0

[32] Liu F, Xu S, Cao L, Chi Y, Zhang T, Xue D. A comparison of $NiMo/Al_2O_3$ catalysts prepared by impregnation and coprecipitation methods for hydrodesulfurization of dibenzothiophene. Journal of Physical Chemistry C. 2007;**111**:7396-7402. DOI: 10.1021/jp068482+

[33] Papadopoulou C, Vakros J, Matralis HK, Voyiatzis GA, Kordulis C. Preparation, characterization, and catalytic activity of $CoMo/Al_2O_3$ catalysts prepared by equilibrium deposition filtration and conventional impregnation techniques. Journal of Colloid and Interface Science. 2004;**274**:159-166. DOI: 10.1016/j.jcis.2003.11.041

[34] Escobar J, De Los Reyes JA, Viveros T. Sol-gel Al_2O_3 structure modification by Ti and Zr addition. A NMR study. Studies in Surface Science and Catalysis. 2000;**143**:547-554. DOI: 10.1016/S0167-2991(00)80696-5

[35] Klimova T. Concerning the interpretation of ^{27}Al MAS-NMR spectra of Mo and NiMo catalysts on Al-containing MCM-41 supports: A reply to the comment by X. Carrier and M. Che on "Ni and Mo interaction with Al-containing MCM-41 support and its effect on the catalytic behavior in DBT hydrodesulfurization" [Appl. Catal. A 240 (2003) 29-40]. Applied Catalysis A: General. 2003;**253**:321-325. DOI: 10.1016/s0926-860x(03)00504-0

[36] Digne M, Sautet P, Raybaud P, Toulhoat H, Artacho E. Structure and stability of aluminum hydroxides: A theoretical study. The Journal of Physical Chemistry B. 2002;**106**: 5155-5162. DOI: 10.1021/jp014182a

[37] Wang JA, Bokhimi X, Morales A, Novaro O. Aluminum local environment and defects in the crystalline structure of sol–gel alumina catalyst. The Journal of Physical Chemistry B. 1999;**103**(2): 299-303. DOI: 10.1021/jp983130r

[38] Cabello CI, Muñoz M, Payen E, Thomas HJ. Influence of cobalt content on the hydrotreatment catalytic activity of $CoMo_6/\gamma\text{-}Al_2O_3$ heteropolyoxomolybdate-based catalyst. Catalysis Letters. 2004;**92**:69-73. DOI: 10.1023/B:CATL.0000011090.71716.6e

[39] Escobar J, Toledo JA, Cortés MA, Mosqueira ML, Pérez V, Ferrat G, et al. Highly active sulfided CoMo catalysts on nano-structured TiO_2. Catalysis Today. 2005;**106**:222-226. DOI: 10.1016/j.cattod.2005.07.136

[40] Dzwigaj S, Louis C, Breysse M, Cattenot M, Belliere V, Geantet C, et al. New generation of titanium dioxide support for hydrodesulfurization. Applied Catalysis B: Environmental. 2003;**41**:181-191. DOI: 10.1016/s0926-3373(02)00210-2

[41] Vuurman MA, Wachs IE. In situ Raman spectroscopy of alumina-supported metal oxide catalysts. The Journal of Physical Chemistry. 1992;**96**:5008-5016. DOI: 10.1021/j100191a051

[42] Guevara-Lara A, Bacaud R, Vrinat M. Highly active NiMo/TiO_2-Al_2O_3 catalysts: Influence of the preparation and the activation conditions on the catalytic activity. Applied Catalysis A: General. 2007;**328**:99-108. DOI: 10.1016/j.apcata.2007.05.028

[43] Christodoulakis A, Heracleous E, Lemonidou AA, Boghosian S. An operando RAMAN study of structure and reactivity of alumina-supported molybdenum oxide catalysts for the oxidative dehydrogenation of ethane. Journal of Catalysis. 2006;**242**:16-25. DOI: 10.1016/j.jcat.2006.05.024

[44] Mojet BL, Coulier L, van Grondelle J, Niemantsverdriet JW, van Santen RA. Potential of UV-RAMAN spectroscopy for characterization of sub-monolayer MoOx model catalysts at ambient pressure. Catalysis Letters. 2004;**96**:1-4. DOI: 10.1023/B:CATL.0000029521.80714.8d

[45] Jia M, Afanasiev P, Vrinat M. The influence of preparation method on the properties of NiMo sulfide catalysts supported on ZrO_2. Applied Catalysis A: General. 2005;**278**:213-221. DOI: 10.1016/j.apcata.2004.09.037

[46] Mazurelle J, Lamonier C, Lancelot C, Edmond P, Pichon C, Guillaume D. Use of the cobalt salt of the heteropolyanion $[Co_2Mo_{10}O_{38}H_4]^{6-}$ for the preparation of CoMo HDS catalysts supported on Al_2O_3, TiO_2 and ZrO_2. Catalysis Today. 2008;**130**:41-49. DOI: 10.1016/j.cattod.2007.07.008

[47] Haber J. Molybdenum compounds in heterogeneous catalysis. In: Braithwaite ER, Haber J, editors. Molybdenum: An Outline of its Chemistry and Uses. 1st ed. Amsterdam: Elsevier Science; 1994. pp. 477-617. ISBN: 9781483290898

Section 3

Technological Applications

Chapter 7

Application of Emulsions and Microemulsions in Enhanced Oil Recovery and Well Stimulation

Mysara E. Mohyaldinn, Anas M. Hassan and Mohammed A. Ayoub

Abstract

Hydrocarbons are produced and transported in a form of mixtures containing oil, gas, and water plus organic and inorganic contaminants. The flow presence of these contaminants (emulsifiers) with the continuous agitation from reservoirs up to surface facilities leads to formation of tight emulsions that need to be dealt with carefully to treat and process them adequately. Emulsions, in the other hand, are sometimes intentionally formed for using in enhanced oil recovery (EOR) and well stimulation. In EOR, emulsions are formed and injected into the reservoirs for the objective of improving both the microscopic displacement efficiency and the macroscopic sweep efficiency, which leads to higher recovery factor. In well stimulation emulsified acids are used during matrix acidizing and acid fracturing to retard acid reaction with rocks, to generate deeper penetration inside the reservoir. Microemulsion is a form of emulsion with less droplet size, and hence higher stability, that occasionally used during EOR and hydraulic fracturing to further improve the reservoir recovery and well production rate. This chapter discusses the application of emulsions and microemulsions in petroleum industry. The chapter discusses emulsions, microemulsions, emulsification processes, application of emulsions and microemulsions in enhanced oil recovery and well stimulations, and ended with conclusions.

Keywords: emulsion, microemulsion, enhanced oil recovery (EOR), microemulsion flooding, microscopic displacement, macroscopic displacement, interfacial tension, surfactant injection, phase behavior, physical stability

1. Introduction

According to Khan et al. [1] "An emulsion may be defined as a biphasic system consisting of two immiscible liquids, one of which (the dispersed phase) is finely and uniformly dispersed as globules throughout the second phase (the continuous phase)." In other words, an emulsion is an insoluble mixture composed of two different types of liquid in which one liquid, i.e., dispersed phase, is spreading in the other liquid, which is in a continuous phase. Emulsions are widely used in our daily lives, examples of the most frequently used emulsions are milk, lotion, and creams. Emulsions in different industries can be formed intentionally to meet required applications, but emulsions occurrence in many cases is undesired and may cause

Emulsions as problems	Emulsions as solutions
Oil treatment More retention time (larger required separator), gas bow-by, formation of rag layer will affect level control system, additional cost (chemical/heating).	Enhance oil recovery The injection of emulsions and/or microemulsion into oil reservoirs is believed to enhance oil recovery due to their favorable rheological, and thermodynamical properties.
Oil transportation Equipment upsets, piping corrosion and scaling, affecting design throughput, single phase correlations become inaccurate when dealing with emulsions.	Well stimulation Emulsified acids are frequently applied to retard acids-rock reaction and hence increase penetration. Microemulsions also are used in hydraulic fracturing

Table 1.
Emulsions as problems and solutions in petroleum industry.

serious problems. In petroleum industry, for instance, emulsions can intentionally be formed for application in some upstream operations such as enhanced oil recovery (EOR), hydraulic fracturing, and emulsified acids. Nevertheless, negative impact due undesired emulsion is also encountered downstream operations. If the produced oil/water streams reach surface facilities in form of stabilized emulsions then more time, money, and technical knowhow are needed to separate the produced streams into distinct phases and make them ready to subsequent treatments. When crude oil is extracted from a well, it frequently leaves the wellhead in a form of mixture containing water phase. This will make it difficult to deal with in surface facilities and transportation areas [2]. **Table 1** introduces where emulsions is considered as problems and solutions in petroleum industry.

2. Emulsion

2.1 Type of emulsion

The two most common types of emulsion are oil-in-water (O/W) emulsion and water-in-oil (W/O) emulsion. There are also multiple emulsions in a complex system. Various types of emulsion can exist in the form of W/O, O/W, O/W/O, and W/O/W, however the most familiar ones are the first two [3]. In oil-in-water emulsions tiny nonpolar droplets of oil spread in a polar water phase [4]. As shown in **Figure 1**, the O/W system appears as a heterogeneous emulsion in which oil is dispersed in a form of bubbles of different sizes in the continuous water phase. In contrary, in water-in-oil emulsions oil forms the primary (continuous) phase in which water droplets disperse with different sizes.

W/O emulsions can further be categorized into stable, entrained, meso-stable, and unstable based on visual appearance, elasticity, and viscosity difference [6]. Nevertheless, in practice only stable and meso-stable states can be considered as real emulsions. The four types of water-in-oil emulsions have been clearly described by Fingas and others [6, 7] as follow:

- Stable emulsions are described as reddish brown semi-solid substances having 70–80% water cut. Experimental examinations conducted by [6] indicated that stable emulsions can remain stable for more than 4 weeks and the stability can extend to more than least 1 year. They also pointed out a dramatical viscosity increase to more than 800 folds after 1 week.

- Meso-stable emulsions the meso-stable water in oil emulsions are viscous liquids and similar in color with the stable emulsions (reddish brown).

Unlike the stable emulsions, the initial value of water content in meso-stable emulsions is in the range 60–65% but the water cut declines to less than 30% after 1 week of formation. The viscosity of meso-stable emulsions also increases with time but the increment rate is not as high as that of stable emulsions [7].

- Entrained emulsions: the entrained emulsions have less water cut and less viscosity increment than that of meso-stable emulsions. The water cut at the date of formation is in the range 40–50% and drops to 28% after 1 week [6].

- Unstable emulsions: contains very little amount of water which almost remain constant with time [7] (**Figure 2**).

In petroleum industry, both W/O and O/W emulsions can be recognized in different locations starting from the reservoir and up to refinery and terminal stations. W/O emulsion is frequently recognized at down-hole and downstream to the choke because at these points the flow stream is mainly hydrocarbon (oil and gas) with lesser volume fraction of water. Due to the continuous turbulence and agitation of the stream while it flows through tubing and passing through wellhead and choke, the water phase breaks down into small droplets and scatters

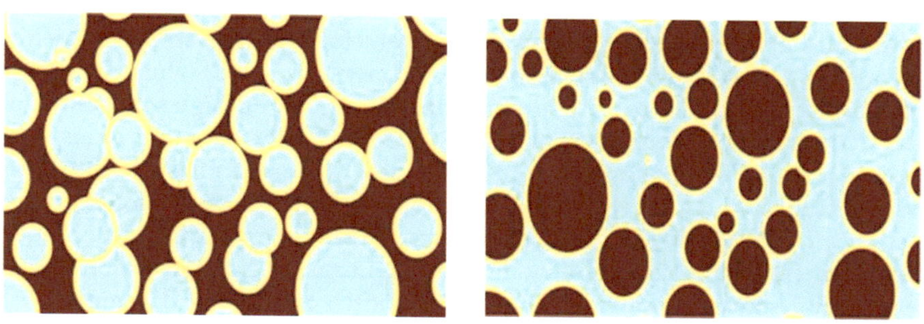

Figure 1.
Water-oil (W/O) emulsion (left) and oil-water (O/W) emulsion (right) [5].

Figure 2.
A photograph of stable emulsion (A) and unstable emulsion (B) [6].

inside the primary oil phase. The presence of impurities in the oil phases, which act as emulsifiers, will increase the possibility of forming W/O emulsion with stability proportional to the degree of turbulence and the concentration of the emulsifiers. O/W emulsion, from the other hand, is more frequently recognized at water outlets downstream to two phase separators and at the inlet of three phase separators and free water knock out (FWKO) facilities. This type of emulsions should be treated properly before it reaches the first pump station for long distance transportation. The main problem arise from flowing of this type of emulsions in pipelines is the corrosion of internal wall of the pipeline due to presence of water. However, presence of emulsions inside pipeline may also be useful as it facilitates smooth transportation of high viscous crude. Because of that heavy and waxy crude oils may intentionally emulsified with water to form (O/W) emulsion.

2.2 Characteristic of emulsion

The main characteristic upon which emulsions are valued is the emulsion stability. Emulsion stability is directly related to the degree of the emulsion tightness and the force of bond between the primary phase and the dispersed phase. High stability is sometimes desired for some applications and processes, whereas it causes problems in others. The applications where high stable emulsions are needed in petroleum industry include enhanced oil recovery (EOR), well stimulation, and heavy/waxy crude transportation. The applications where high stable emulsions are problematic are downstream operations. Highly stable emulsion causes a lot of problems in separation and treatment processes where longer time and more cost are needed to ensure efficient separation and high sale value for the treated hydrocarbons. Emulsion stability is affected by the characteristics of the two immiscible phases (the continuous and dispersed phases), the degree of the agitation to which the mixture is subjected, and the concentration and type of emulsifiers. Thermal stability can be measured in lab using different techniques or can be quantified using the stability indexed as discussed before. The simplest and more accessible measurement technique is the bottle test. A small amount of the emulsion is placed in a scaled bottle and the separated amount of the dispersed phase is noticed over time (**Figure 3**).

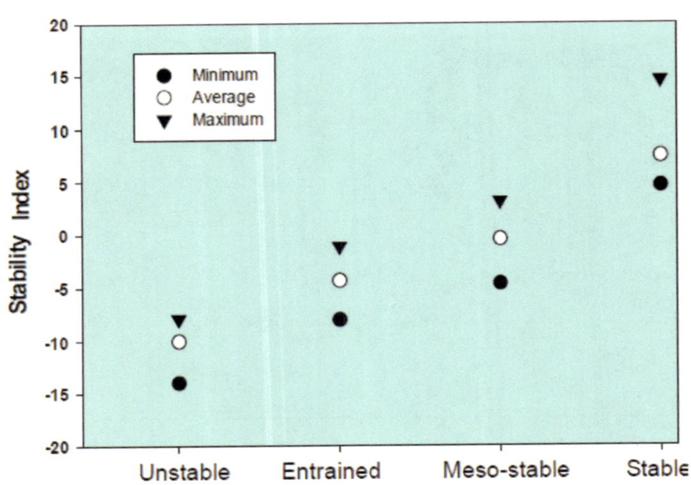

Figure 3.
Water-oil W/O emulsion identification based on stability index [6].

2.2.1 Effect of rheology on emulsion stability

The effect of different flow properties on emulsion has been studied by many researchers [8–10]. The rheology of an emulsion involves the study of the flow of the emulsion itself. Therefore, it concerns with the viscosity of the emulsion. Some of the rheological properties are temperature, dispersed phase volume fraction, the inclusion of surfactant in emulsion and the chemical constituent that exists in each section. All of these properties affect the viscosity of emulsion which is usually measured. It was also revealed that the viscosity becomes substantially lesser as the volume fraction of water was lowered. Even more so, the viscosity is also firmly influenced by temperature [11]. The temperature when it is intensifying at a certain degree it will cause a cracking between the bond with solid components as it will not allow the components which contain in crude oil to have the opportunity to combine and agglomerated. Eventually, less viscous emulsion in obtained. Additionally, other factors that affect the rheology are the shear rate. Greater shear rates result in lower viscosity as there will be less flow ability interruption in the emulsion [2]. Although shear rate is considered to affect the rheology, Keleşoğlu et al. [12] proclaimed that shear rate does not apply to an emulsion with a volume fraction that is less dispersed. Out of all the factors states, temperature is one of the significant parameter that can highly affect the viscosity of emulsion.

- Emulsions are encountered at different steps of petroleum engineering related activities, starting from the moment the hydrocarbons are contained inside the reservoirs and up to separation and treatment of oil and gas.

- The emulsions are normally considered as problematic when thinking about it from the oil processing and treatment point of view. In this case, an undesirably formed emulsion leads to slowing down oil/gas/water separation and process and, by such, increasing the cost of processing and treatment. In such a case, the emulsion need to be destabilized and broken by heating and/or using chemical demulsifiers,

- Emulsions, from the other hand, may intentionally be formed and stabilized for application in enhanced oil recovery and well stimulations.

- Emulsion stability is affected by (1) the characteristics of the two immiscible phases (the continuous and dispersed phases), (2) the degree of the agitation to which the mixture is subjected, and (3) the concentration and type of emulsifiers.

- Microemulsions are generally composed of hydrocarbons, surfactants/co-surfactants and brine.

- Macro/microemulsions based enhanced oil recovery improve both the microscopic displacement efficiency and the macroscopic sweep efficiency, thus leads to higher recovery factor.

- In enhanced oil recovery, the microemulsion flooding displays the unique properties of microemulsion systems, such as high viscosity and the ability to induce low interfacial tension, increasing oil extraction efficiency.

- In well stimulation emulsified acids are used during matrix acidizing and acid fracturing to retard acid reaction with rocks, to generate deeper penetration inside the reservoir.

3. Emulsification

Emulsion is usually formed by the action of mechanical mixing with the existence of a surfactant molecule. The mechanical mixing action can generate a turbulence effect and break down the two immiscible phases, while surfactant is contributing in building an adhesion force on the surface between the droplets of the two phases (see **Figure 4**). This process of the formation of an emulsion is called emulsification. There are two types of emulsification, spontaneous emulsification and self-emulsification. The spontaneous emulsification is the emulsification process where there is no involvement of external heat or physical action or energy that can affect the emulsification of the two immiscible liquids. In the spontaneous emulsification, the two phases usually take a significant amount of time to completely become an emulsion. In the self-emulsification, the complete emulsification occurs when appropriate surfactants are used [14]. Emulsification can be observed in many activities and processes ranging from simple events people frequently do in daily life to more complicated and sophisticated processes performed in industry. Producing mayonnaise is an example of the emulsification processes where egg yolks and oil are mixed and emulsified by stirring action to form a thick emulsion in which the egg yolk acts as a continuous phase while the oil is the dispersed phase. Similarly, mixing water in oil can generate a water-in-oil or oil-in-water emulsion, depending on which phase is predominate. In this case, however, an emulsified material such as asphaltenes is needed to stabilize the emulsion. Fingas and Fieldhouse [6] mentioned that asphaltenes are the main factor that cause (W/O) emulsion stabilization as it has the ability to build a rigid cross-linked and elastic films. Besides asphaltenes, resins can also act as stabilizing agents. As outlined by Fingas [7], although resins normally assist the asphaltenes by acting as a solvent that can stabilize as the asphaltenes migrates. This shows that asphaltenes and resins can be considered as the main emulsifiers that exist naturally in crude oils. In addition, to asphaltenes and resins, other organic and inorganic stabilizers frequently found in crude oils include waxes and clays [15]. Crude oil does not always appear as a stable emulsion as there is certain phenomenon that continuously destabilizes it. Examples of such phenomenon are flocculation, creaming, and coalescence. Flocculation is the process of the combination of small emulsion droplets each other due to excess surfactants in the continuous phase. When the emulsion is shaken, however, these particles can disperse again in the medium.

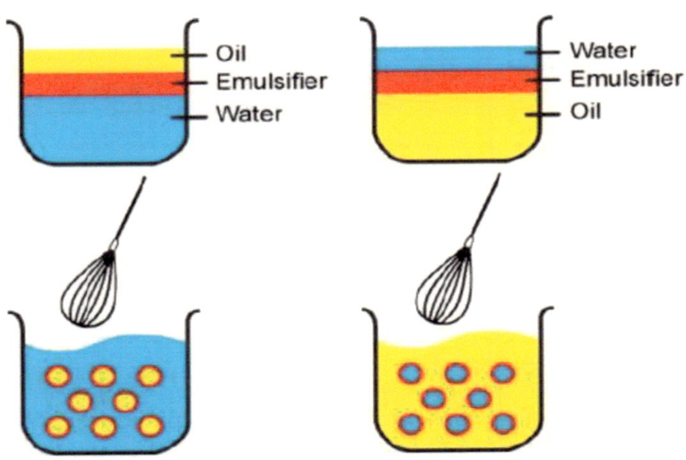

Figure 4.
Main basic types of microemulsions; the three basic types of microemulsions are direct (oil dispersed in water, o/w), reversed (water dispersed in oil, w/o) and bi-continuous [13].

4. Microemulsions

4.1 Type and structure of microemulsion

Microemulsion structure has a key role in the different physicochemical properties of the applied fields. The specific structures of the microemulsions have been extensively studied by many scientist researchers [16–22]. The three basic types of microemulsions are direct (oil dispersed in water, o/w), reversed (water dispersed in oil, w/o) and bi-continuous. Like multiple emulsion, sometimes multiple microemulsion is also possible. In this type, another layer is formed outside the o/w or w/o microemulsions. The schematic diagram of the basic three types of microemulsions is revealed in **Figure 5**. Microemulsion structure depends on salinity, water content, co-surfactant concentration and surfactant concentration. At higher water content, the microemulsion would be a water-external system, with oil solubilized in the cores of the micelles. Although the mixtures remain single phase and thermodynamically stable, the microemulsion structure changes through a series of intermediate states [23]. The structures of these intermediate states are not well known. However, the solutions are thermodynamically stable and isotropic. Salinity also can reverse the structure of the microemulsion. As salinity increases, the direct microemulsion changes to reverse microemulsions. At low salinity, the system remains in water-external phase, but with increasing salinity the system separates into an oil-external microemulsion.

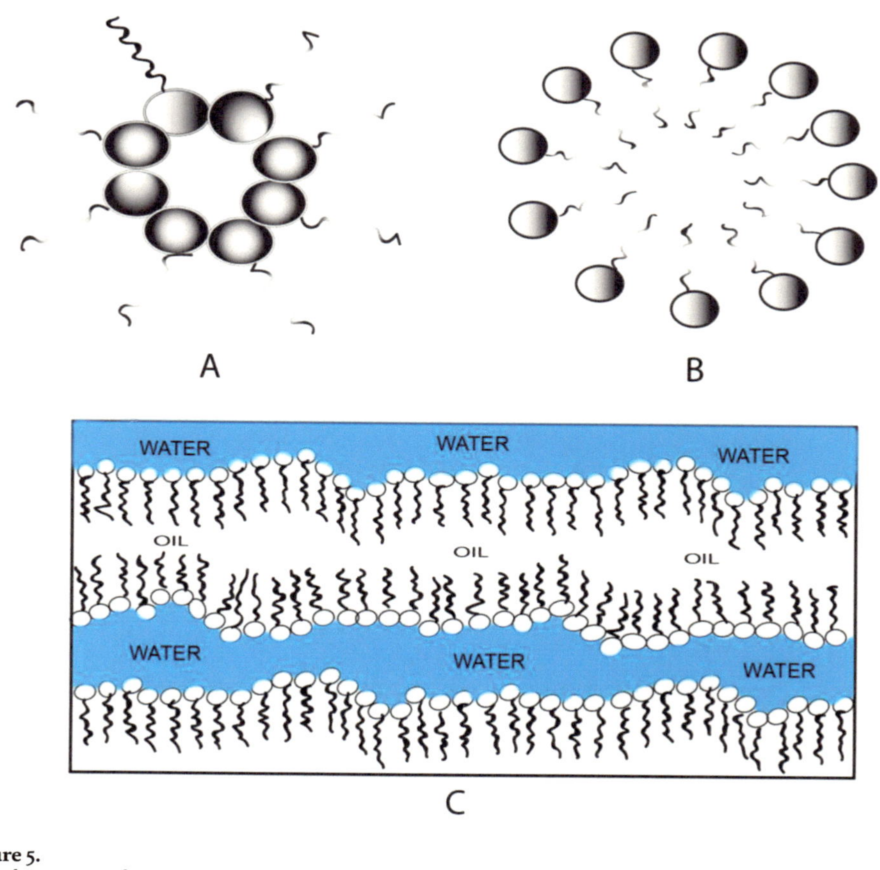

Figure 5.
Main basic types of microemulsions; the three basic types of microemulsions are (A) reversed (water dispersed in oil, w/o), (B) direct (oil dispersed in water, o/w), and (C) bi-continuous [16].

The term "microemulsions" has been introduced for the first time by Schulman et al. [24]. The term was used to describe what is called transparent solutions in a model four component system consisting of water, hydrocarbon, surfactant, and co-surfactant. Microemulsions can be experimentally investigated and described using different characterization apparatus such as low-angle X-ray diffraction [25] and viscosity measurements [26]. Microemulsions [27–29] can also be identified and described using phase mapping which is a tetrahedron shape showing the components of the microemulsions (i.e., water, hydrocarbon, ionic surfactant, and co-surfactant) in the four corners of the tetrahedron, as shown in **Figure 6**.

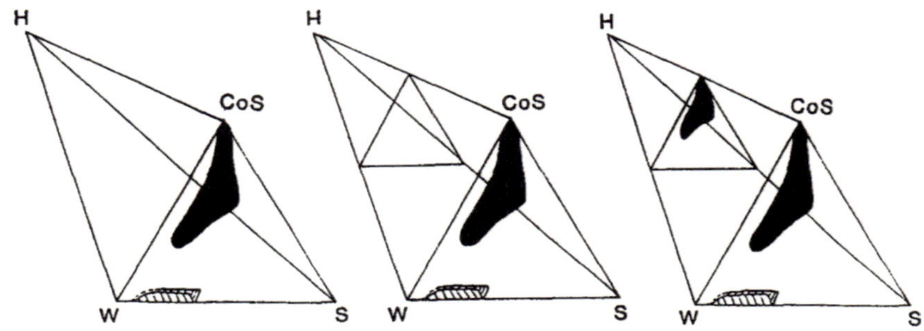

Figure 6.
The four component of microemulsion system water (W), hydrocarbon (H), ionic surfactant (S) and co-surfactant (CoS) [30].

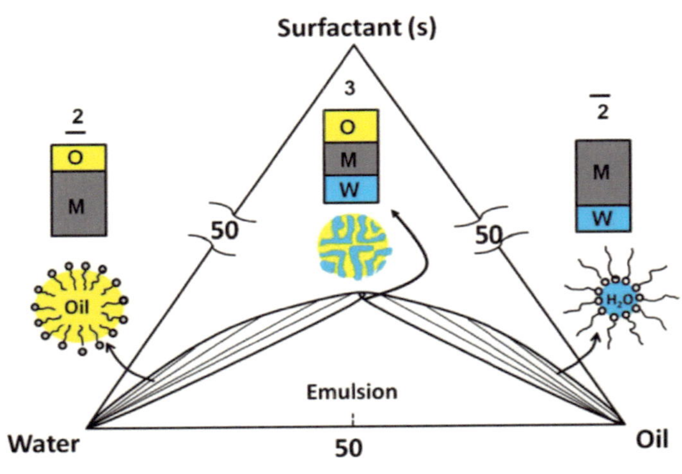

Figure 7.
Typical water-surfactant-oil microemulsions as depicted by the ternary phase diagram [31, 32].

Similar to emulsion classifications, microemulsions are classified into three types as follow:

- Oil-in-water (O/W) microemulsion: in this type, i.e., the water phase predominates the oil phase, the oil volume fraction is smaller than the water volume fraction.
- Water-in-oil (W/O) microemulsion: in contrary to the O/W microemulsions, in this type the oil phase volume fraction predominates the water phase.

- Bi-continuous microemulsion: also referred to as "sponge-like" microemulsions. In this type the volume fractions of oil and water re almost similar and both can be considered as continuous, while the surfactant molecules is forming rapidly fluctuating interfaces [27, 28].

The phase diagram for a typical there-component surfactant-water-oil is shown in **Figure** 7 [31, 32].

5. Demulsification

Demulsification play an important key in separating water and salts from crude oil which is vital during industrial procedure [33]. Since crude oil consists of impurities and if it is not removed, thus it will cause fouling and major corrosion to the equipment in the industry. As a matter of fact, this process can only be executed when demulsifiers are utilized. This chemical component is capable to inflict the coalescence of water droplets by forming film drainage and elevate the surface activity as the gradient is inversed. Thus, it means that demulsifier can undeniably change the interface physical characteristics. When demulsifiers are added to a dilute emulsion of low concentration, a specific process call adsorption takes place and it will adsorb the emulsion particles and situate it on the surface in forms of droplets. Furthermore, the demulsifier contain organic particles and it will locate and adsorb the dispersed phase in such a way that the non-polar part will be in the crude oil while the polar part will maintain in water [34]. Demulsifiers are also known as a non-ionic that consists of two separate parts of hydrophobic and hydrophilic. The hydrophobic group contains oxypropylenes, alkyls or alkyl phenols whereas the hydrophilic group contains amine groups, carboxyl, hydroxyl or oxyethylene. There are also several methods in separating the emulsions which involves separation in terms of chemical, electrical and mechanical [34]. Usually, it is observed in the industry, the use of chemical method is the most common. The coalescence of (W/O) emulsion is maximized and the protective film breaks with the use of chemical demulsifier [35]. As a result, researchers are looking for a way to accelerate the demulsification process effectively. Several methods were applied, for example, the use of microwave energy, adding Janus magnetic microparticle in demulsifier and others. Based on the research conducted by Martínez-Palou et al. [3]. It was found that the rate of separation of (W/O) emulsion were highly efficient when microwave power was applied compared to the use of chemical non-ionic surfactants and also the solubility of surfactants in water decreases with the aid of a saline solution like seawater. Conversely, the application of the Janus magnetic submicronic particles or P(MMAAA-DVB)/Fe_3O_4 was very promising as it exhibits a high rate of coalescence whereby the water droplets are settled magnetically through the attraction of an outer magnetic field and it is also recyclable, hence lowering the cost in the petroleum industry [36]. To enhance the demulsification efficiency, the factors that can affect it are the molecular weight, hydrophile lipophile balance (HLB), concentration, water content, temperature and asphaltene content. Concerning the molecular weight, the lower the molecular weight of the demulsifier, the higher the rate of partitioning the emulsion as an elevation in the molecular weight causes a difficulty for the demulsifier to spread in the highly thick or viscous crude oil. Beside, a rise in the amount of HLB, concentration and water content shows an increasing rate in the efficiency of this process. However, the asphaltenes which acts as stabilizing agents can prevent the demulsification of crude oil because when the content becomes higher, it will result a dense film that protects the emulsion. To summarize, the complete separation of (O/W) emulsion can be attain based on two prime factors which is the temperature and the quantity of asphaltenes [32, 37].

6. Applications of emulsion/microemulsion in oil industry

6.1 Background

It is to be expected that the future energy demand will be met by a global energy mix that is undergoing a transition from the current dominance of fossil fuels to a more balanced distribution of energy sources. After conventional waterflood processes, the part of the oil in the reservoir remains as a discontinuous phase in the form of oil globules trapped by capillary forces and is likely to be around 30% of the original oil in place (OOIP) [38], whereas another 40% is bypassed by the water. However, technically it is possible to improve this recovery efficiency by applying enhanced oil recovery (EOR) processes. Traditionally, oil recovery operations comprise three stages: primary, secondary, and tertiary. These stages describe the production from a reservoir in chronological manner. During primary recovery, which starts right from the earliest stages of production, the energy for displacing the oil to the production wells results from the use of natural energy present in the reservoir [39]. Continual withdrawal of the fluids from the reservoir, results in a decrease of the supporting energy. Consequently, it reaches a stage whereby further removal of fluids approaches the limits of a profitable operation. Here, it requires an intervention for increasing the reservoir energy and thus fostering production. This implies the initiation of the second stage, which entails the application of secondary recovery methods. The secondary recovery method results in the augmentation of the natural energy of the reservoir. One of the most popular secondary recovery methods is waterflooding. The estimations showed that the unproduced residual oil reserves from primary and secondary oil production methods are about two to three trillion barrels worldwide [40]. Given the tremendous amount of the unrecovered oil, the introduction of new intervention techniques is crucial. This entails the application of methods for tertiary recovery or enhanced oil recovery (EOR). Typically, the methods consist of injection of (nonreservoir) gases [40, 41], liquid chemicals and/or the use of thermal energy. This work focuses on the use of emulsions and/or microemulsions as effective method to improve oil recovery in the field. Previous experiences clearly indicate that emulsion and/or microemulsion-based EOR methods are an ideal, as well as a feasible alternative that can effectively recover this enormous resource base [42, 43].

6.2 Applications in enhanced oil recovery (EOR)

In oil and gas industry, the approach to emulsion and/or microemulsion preparation has associated with the application of energy to a mixture of oil, water, and emulsifier. The emulsifier (i.e., surfactants or amphiphilic proteins) acts to stabilize the interfacial layer between the continuous and dispersed phase which has been generated through the addition of energy to the system. The injection of emulsions and/or microemulsion into oil reservoirs has been acknowledge as a potential tool for oil recovery due to the possibility of rheological, and thermodynamical properties of emulsions [44]. As reviewed by Muggeridge et al. [45] main objective of oil recovery is to improve both the microscopic displacement efficiency and the macroscopic sweep efficiency of the reservoir. The former regards the removal of oil at the pore level, while the latter is targeted at removing oil at the larger scale of the medium by avoiding oil trapping, a feature related to reservoir geological conformation and surface chemistry.

6.2.1 Macro- and microemulsion flooding

The use of microemulsions is of high interest in many aspects of crude oil exploitation, especially in enhanced oil recovery (EOR). Macro- and microemulsion flooding is an efficient EOR recovery method due to its high extraction efficiency by reducing the oil-water interfacial tension [42]. A macroemulsion is a thermodynamics unstable heterogeneous mixture of oil and water with either oil droplet in water (an oil/water emulsion) or water droplets in oil (a water/oil emulsion). The droplets are stabilized by surfactants that absorb at the oil-water inter-phase, which makes that the interfaces are charged preventing the droplets to collide. On other hand, microemulsions are transparent homogeneous mixtures of hydrocarbons and water with large amounts of surfactants [24]. The microemulsions have recently been introduced in enhanced oil recovery processes in which chemicals, especially surfactants, are used to recover the oil from natural oil reservoirs. Since the discovery of microemulsions, they have attained increasing significance both in basic research studies and in the oil fields. In spite of intensive research on microemulsions, the theory behind understanding mechanisms (i.e., predictive power) for microemulsions is still lacking [43, 46, 47]. In addition, many difficulties are encountered in creating a suitable microemulsion film with temperature gradients required large, many ionic surfactants precipitating when contacted with brine, and most non-ionic surfactants unsuitable. Other issues include adsorption of microemulsion components on rocks, and varying salinities and temperatures of the oil reservoirs. Over the past three decades, however, there has been sufficiently great progress made on the recovery of residual oil in particular chemical based enhanced oil recovery with microemulsions. Babadagli [48] has written a review about improvement of mature oil fields. According to his analysis, the most common chemical-based enhanced oil recovery method is the surfactant solution injection due to its relatively lower cost when compared to micellar or microemulsion injection. The way by which the injected chemical, in chemical based enhanced oil recovery (EOR), is a significant parameter for optimization of the EOR-method. Continuous injection of a chemical solution leads to increase operation costs and/or reduce the amount of treated material. Taking into account, injection of chemical solution considering the porous volume number (PV) is required in any efficient enhanced oil recovery (EOR) process. Thomas and co-workers (in: [48]) injected porous volumes (PV) of microemulsion in sandstone rock containing 35% of residual oil, noticing a linear relationship between the values of injected pore volume (PV) and the oil recovery. The Results of the study of Thomas and co-workers showed a 45% residual oil recovery when injecting around 10 pore volume (PV) of microemulsion. Santanna et al. [42] studied the application of different types of microemulsion for chemical based enhanced oil recovery (EOR), including different types of surfactants [42, 49, 50].

6.2.2 Surfactant microemulsion flooding

Microemulsions are generally composed of hydrocarbons, surfactants/co-surfactants and brine. Surfactants are considered to be the principal constituents of microemulsions and are adsorbed at the interface rather than in the bulk phase. Surfactants are classified into four groups based on the charge of the head group such as anionic, cationic, non-ionic and zwitterionic. Anionic surfactants such as sodium dodecylsulfate (SDS) are negatively charged in nature, but a small cation sodium ion occupies the counterpart. Anionic surfactants are most widely used in oil recovery process. Their adsorption phenomena in sandstone and carbonate are different. Their adsorption in sandstone is relatively lower than that of carbonate [49–51].

6.2.3 Mechanism of surfactant microemulsion flooding

In microemulsion flooding, the reservoir is flooded with water containing a small percentage of surfactant and other additives such as hydrocarbon, medium-chain alcohol and brine. The surfactant plays a key role in forming the exact type of microemulsion that reduces the interfacial tension of the target oil [29, 52]. This is critical to both mobilize oil and enable it to escape from the reservoir rock. In general, whenever a waterflood has been successful, microemulsion injection will be applicable, while in many cases where water injection has failed due to its poor mobility relationships, microemulsion flooding can still be successful mainly because the required mobility control.

6.2.4 Interfacial tension reduction

In enhanced oil recovery, the microemulsion flooding displays the unique properties of microemulsion systems, such as high viscosity and the ability to induce low interfacial tension, increasing oil extraction efficiency. [42] According to Austad and Strand [53, 54], very low interfacial tensions may be reached with microemulsion systems. As stated by Gurgel et al. [55] microemulsions are potential candidates in enhanced oil recovery, especially because of its ultra-low interfacial tension values, attained between the contacting oil and water microphases that form them. Under such circumstances, microemulsions flow more easily through the porous medium, which enhance oil extraction performance rates. Babadagli [48] has written a review about improvement of mature oil fields. According to his review, the most common chemical injection technique, as an enhanced oil recovery method, is the surfactant solution injection due to its relatively lower cost when compared to microemulsion injection. The way by which the fluid is injected, when the chemical method is applied, is an important parameter for optimization of the technique. Continuous injection of a chemical solution may increase operation costs and reduce the amount of treated material. In view of this, injection of chemical solution considering the porous volume number (PV) is required in any efficient recovery process. Thomas et al. (in: [48]) injected porous volumes of microemulsion in sandstone plugs containing 35% of residual oil, observing a linear relationship between the values of injected PV and the oil recovery. Results typically showed a 45% residual oil recovery when injecting 10 PV of microemulsion. Santanna and co-workers [42] studied the application of different types of microemulsion for enhanced oil recovery, one was prepared with a commercial surfactant (MCS), and another contained a surfactant synthesized in laboratory (MLS). The experiments consisted of the injection of fluids into cylindrical plug samples. During the microemulsion flooding, samples were collected as a function of time and the mass of oil recovered by the microemulsion was determined. The chemicals used to prepare the microemulsion systems were commercial anionic surfactant (soap-sodium salt) obtained from fatty acids; anionic surfactant (soap–sodium salt) synthesized in laboratory, extract from fatty acids (100 wt.% of vegetable oil containing 12 carbon atoms); isoamyl alcohol; pine oil; and distilled water. From the results obtained, one could conclude that the use of microemulsion prepared with the commercial MCS allowed for recovery indexes as high as 87.5%, whilst the use of the MLS microemulsion permitted recovery indexes as high as 78.7%. This was because the difference in microemulsion viscosities, corroborated by the fact that the MCS-based microemulsion (32 cP viscosity) could recover more oil than the MLS-based microemulsion (27 cP viscosity).

6.3 Emulsion and microemulsion applications in well stimulation

Matrix acidizing is a well stimulation technique used to eliminate formation damage and/or increase permeability in sandstone and carbonate reservoir. This eventually leads to the improvement of the reservoir inflow performance, which will turn in enhancing wells productivity [56]. Matrix acidizing is performed by injecting acid solution with a prescribed concentration into the formation from surface. The typical acid used for carbonate reservoirs is hydrochloric acid with concentration of 15%, whereas for sandstone reservoirs a mixture of hydrofluoric/hydrochloric acid solution with a concentration 3% HF/12% HCl is used. The efficiency of matrix acidizing process is highly dependent on the distance the acid can penetrate inside the reservoir, which is inversely proportional to the rate of reaction between the acid and the rock. Fast acid reaction yields less penetration distance inside the formation, and hence, the acidizing operation will get less efficient. Therefore, different methods are implemented to retard the acid-rock reaction. One way to do so is by emulsifying the acid solution with a hydrocarbon oil (diesel oil or xylene) with the assistance of an emulsifying agent. This process results in producing an emulsifying acid fluid that more favorable than the ordinary acids because the hydrocarbon oil phase provides a diffusion barrier causing the slow release of acid and deeper penetration [57]. The emulsified acid is considered efficient if it can slow down the reaction rate (i.e., retard the acid) and generate a successful stimulation operation (expressed as increase in the rock permeability). Both effects can be assessed experimentally using core samples from the target rock. The acid retarding effect can be investigated by comparing the solubility of the rock in the blank acid with that in the emulsified acid. Assessment of emulsified acid retarding efficiency (acid solubility test): the acid solubility test is conducted to measure the amount of rock soluble by using the emulsified acid. Mohsin and his co-workers [56] conducted acid solubility test for an Indiana Limestone core sample in hydrochloric acid oil emulsified acid at ambient and 70°C. Approximately 1 g powder of the rock placed in a volume of 150 cc of the emulsified acid for a period of 60 min without stirring at the desired temperature. Hydrochloric acid solution with 15% concentration has been used to formulate three types of emulsified acids each with different type of oil phase (diesel oil, palm oil, and *Jatropha* oil). The solubility value, expressed in percentage, was calculated for each emulsified acid using the following formula:

$$SA\,\% = W1 - W2/(W1 \times 100) \tag{1}$$

The solubility values at temperature of 70°C are shown in **Table 2**.

6.3.1 Assessment of stimulation efficiency

The stimulation efficiency is investigated experimentally using core flooding lab. A core sample from the stimulated rock is placed inside the core holder of the

Emulsion type	Initial weight (W1), g	Final weight (W2), g	Average solubility (70°C)
Jatropha oil-based	1	0.050033	94.99
Diesel oil-based	1	0.0285	97.15
Palm oil-based	1	0.02076	97.924

Table 2.
Emulsified acids with continuous phase of three types of oil [56].

Figure 8.
The cores used for investigating the diesel-based and Jatropha-based emulsified acids [56].

Figure 9.
Filtrate from Jatropha oil-based emulsion (on left side) and filtrate from diesel oil-based emulsion (on right side) [56].

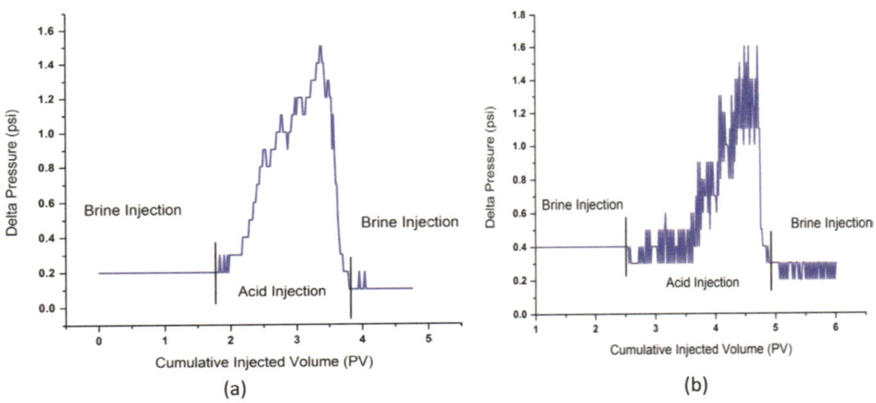

Figure 10.
Pressure drop profiles by the end of the acidizing using (a) diesel based emulsified acid, and (b) Jatropha based emulsified acid [56].

core flooding lab and the reservoir condition can be simulated by setting the pressure and temperature of the core flooding lab at values similar to the real conditions of the reservoir. The core flooding equipment FES 350 has been used by Mohsin and his co-workers [56] to investigate the stimulation efficiency of the diesel-based and *Jatropha*-based emulsified fluids described in **Table 2** at a temperature of 7°C, a back pressure of 700 psi, confining pressure of 1300 psi, and injection rate of 0.5 cc/min. Two carbonate cores, shown in **Figure 8**, were immersed first with brine to determine permeability of the core by recording the differential pressure points using Darcy's law, as the fluid moves across the core. The 15 wt% HCl emulsified acid was then injected until pressure drop was observed indicating the breakthrough. The cores were then flushed with brine to remove stimulation fluid and was cleaned, dried and was measured for weight, porosity and permeability. **Figure 9** shows filtrate from *Jatropha* oil-based emulsion (on left side) and filtrate from diesel oil-based emulsion (on right side).

Figure 10 shows pressure drop profiles of the cores shown in **Figure 8** by the end of the acidizing with the *Jatropha* oil-based (a) and the diesel oil-based (b) emulsified acids. Both the two pressure profiles figures indicate reaching the end of the core by the acidizing medium while creating the conduit and both cores achieved almost similar pressure drop. However, diesel laced acid achieved slightly higher pressure drop (1.6 psi), whereas *Jatropha* oil-based emulsified acid achieved one peak pressure drop at 1.5 psi. This stipulates higher consumption of the diesel-based acid and possibility of increase residue production due to acid reactivity.

7. Conclusions

- An emulsion may be defined as a biphasic system consisting of two immiscible liquids, one of which (the dispersed phase) is finely and uniformly dispersed as globules throughout the second phase (the continuous phase).

- The main characteristic upon which emulsions are valued is the emulsion stability.

- Emulsion stability is directly related to the degree of the emulsion tightness and the force of bond between the primary phase and the dispersed phase.

- Emulsion stability is affected by (1) the characteristics of the two immiscible phases (the continuous and dispersed phases), (2) the degree of the agitation to which the mixture is subjected, and (3) the concentration and type of emulsifiers.

- Microemulsions are generally composed of hydrocarbons, surfactants/co-surfactants and brine.

- It is to be expected that the future energy demand will be met by a global energy mix that is undergoing a transition from the current dominance of fossil fuels to a more balanced distribution of energy sources.

- Macro/microemulsions based enhanced oil recovery improve both the microscopic displacement efficiency and the macroscopic sweep efficiency, thus leads to higher recovery factor.

- In enhanced oil recovery, the microemulsion flooding displays the unique properties of microemulsion systems, such as high viscosity and the ability to induce low interfacial tension, increasing oil extraction efficiency.

- In well stimulation emulsified acids are used during matrix acidizing and acid fracturing to retard acid reaction with rocks, to generate deeper penetration inside the reservoir.

Author details

Mysara E. Mohyaldinn*, Anas M. Hassan and Mohammed A. Ayoub
Department of Petroleum Engineering, Universiti Teknologi PETRONAS (UTP), Teronoh, Perak, Malaysia

*Address all correspondence to: mysara.eissa@utp.edu.my

References

[1] Khan BA, Akhtar N, Khan HMS, Waseem K, Mahmood T, Rasul A, et al. Basics of pharmaceutical emulsions: A review. African Journal of Pharmacy and Pharmacology. 2011;**5**(25):2715-2725

[2] Hasan SW, Ghannam MT, Esmail N. Heavy crude oil viscosity reduction and rheology for pipeline transportation. Fuel. 2010;**89**(5):1095-1100

[3] Martínez-Palou R, Cerón-Camacho R, Chávez B, Vallejo AA, Villanueva-Negrete D, Castellanos J, et al. Demulsification of heavy crude oil-in-water emulsions: A comparative study between microwave and thermal heating. Fuel. 2013;**113**:407-414

[4] Becerra HQ, Retamoso C, Macdonald DD. The corrosion of carbon steel in oil-in-water emulsions under controlled hydrodynamic conditions. Corrosion Science. 2000;**42**(3):561-575

[5] Uma A-B, Saaid IBM, AdebayoSulaimon A, Pilus RBM. A review of petroleum emulsions and recent progress on waterin-crude oil emulsions stabilized by natural surfactants and solids. Journal of Petroleum Science and Engineering. June 2018;**165**:673-690

[6] Fingas M, Fieldhouse B. Studies on crude oil and petroleum product emulsions: Water resolution and rheology. Colloids and Surfaces A: Physicochemical and Engineering Aspects. 2009;**333**(1-3):67-81

[7] Wong SF, Lim JS, Dol SS. Crude oil emulsion: A review on formation, classification and stability of water-in-oil emulsions. Journal of Petroleum Science and Engineering. 2015;**135**:498-504

[8] Floury J, Desrumaux A, Lardieres J. Effect of high-pressure homogenization on droplet size distributions and rheological properties of model oil-in-water emulsions. Innovative Food Science & Emerging Technologies. 2000;**1**(2):127-134

[9] Pal R. Effect of droplet size on the rheology of emulsions. AICHE Journal. 1996;**42**(11):3181-3190

[10] Mohammed RA, Bailey AI, Luckham PF, Taylor SE. The effect of demulsifiers on the interfacial rheology and emulsion stability of water-in-crude oil emulsions. Colloids and Surfaces A: Physicochemical and Engineering Aspects. 1994;**91**:129-139

[11] Akbari S, Nour AH, Jamari SS, Fayaz F. Rheology and stability mechanism of water-in-crude oil emulsions stabilized by span 83. ARPN Journal of Engineering and Applied Sciences. 2006;**11**(4)

[12] Keleşoğlu S, Pettersen BH, Sjöblom J. Flow properties of water-in north sea heavy crude oil emulsions. Journal of Petroleum Science and Engineering. 2012;**100**:14-23

[13] Madaan V, Chanana K, Kataria MK, Bilandi A. Emulsion technology and recent trends in emulsion applications. International Research Journal of Pharmacy. 2014;**5**(7):533-542

[14] López-Montilla JC, Herrera-Morales PE, Pandey S, Shah DO. Spontaneous emulsification: Mechanisms, physicochemical aspects, modeling, and applications. Journal of Dispersion Science and Technology. 2002;**23**(1-3):219-268

[15] Bhardwaj A, Hartland S. Studies on build-up of interfacial film at the crude oil/water interface. Journal of Dispersion Science and Technology. 1998;**19**(4):465-473

[16] Bera A, Mandal A. Microemulsions: A novel approach to enhanced oil

recovery: A review. Journal of Petroleum Exploration and Production Technology. 2015;**5**(3):255-268

[17] Li X, He G, Zheng W, Xiao G. Study on conductivity property and microstructure of Triton X-100/alkanol/n-heptane/water microemulsion. Colloids and Surfaces A: Physicochemical and Engineering Aspects. 2010;**360**(1-3):150-158

[18] Torino E, Reverchon E, Johnston KP. Carbon dioxide/water, water/carbon dioxide emulsions and double emulsions stabilized with a nonionic biocompatible surfactant. Journal of Colloid and Interface Science. 2010;**348**(2):469-478

[19] Zemb T. Flexibility, persistence length and bicontinuous microstructures in microemulsions. Comptes Rendus Chimie. 2009;**12**(1-2):218-224

[20] Mathew DS, Juang R-S. An overview of the structure and magnetism of spinel ferrite nanoparticles and their synthesis in microemulsions. Chemical Engineering Journal. 2007;**129**(1-3):51-65

[21] Ben Azouz I, Ober R, Nakache E, Williams CE. A small angle X-ray scattering investigation of the structure of a ternary water-in-oil microemulsion. Colloids and Surfaces. 1992;**69**(2-3):87-97

[22] Guest D, Langevin D. Light scattering study of a multiphase microemulsion system. Journal of Colloid and Interface Science. 1986;**112**(1):208-220

[23] Bartscherer KA, Minier M, Renon H. Microemulsions in compressible fluids: A review. Fluid Phase Equilibria. 1995;**107**(1):93-150

[24] Schulman JH, Stoeckenius W, Prince LM. Mechanism of formation and structure of micro emulsions by electron microscopy. The Journal of Physical Chemistry. 1959;**63**(10):1677-1680

[25] Stevens MP. Polymer Chemistry. New York: Oxford Univ. Press; 1990

[26] Georges J, Chen JW. Microemulsions studies: Correlation between viscosity, electrical conductivity and electrochemical and fluorescent probe measurements. Colloid and Polymer Science. 1986;**264**(10):896-902

[27] Binks BP, Fletcher PDI, Tian L. Influence of nanoparticle addition to winsor surfactant microemulsion systems. Colloids and Surfaces A: Physicochemical and Engineering Aspects. 2010;**363**(1-3):8-15

[28] Ajith S, Rakshit K. Studies of mixed surfactant microemulsion systems: Brij 35 with Tween 20 and sodium dodecyl sulfate. The Journal of Physical Chemistry. 1995;**99**(40):14778-14783

[29] Shinoda K, Friberg S. Emulsions and Solubilization. New York: Wiley-Interscience; 1986

[30] Sjöblom J, Lindberg R, Friberg SE. Microemulsions: Phase equilibria characterization, structures, applications and chemical reactions. Advances in Colloid and Interface Science. 1996;**65**:125-287

[31] Yaghmur A, Aserin A, Garti N. Phase behavior of microemulsions based on food-grade nonionic surfactants: Effect of polyols and short-chain alcohols. Colloids and Surfaces A: Physicochemical and Engineering Aspects. 2002;**209**(1):71-81

[32] Safran SA, Turkevich LA, Al-Sabagh AM, El-Din MRN, Morsi RE, Elsabee MZ. Styrene-maleic anhydride copolymer esters as flow improvers of waxy crude oil. Journal of Petroleum Science and Engineering. 2009;**65**(3-4):139-146

[33] Krawczyk MA, Wasan DT, Shetty C. Chemical demulsification of petroleum emulsions using oil-soluble demulsifiers. Industrial & Engineering Chemistry Research. 1991;**30**(2):367-375

[34] Al-Sabagh AM, Kandile NG, El-Ghazawy RA, Noor ElDin MR. Synthesis and evaluation of some new demulsifiers based on bisphenols for treating water-in-crude oil emulsions. Egyptian Journal of Petroleum. 2011;**20**(2):67-77

[35] Mosayebi A, Abedini R. Using demulsifiers for phase breaking of water/oil emulsion. Petroleum and Coal. 2013;**55**(1):26-30

[36] Ali N, Zhang B, Zhang H, Li W, Zaman W, Tian L, et al. Novel janus magnetic micro particle synthesis and its applications as a demulsifier for breaking heavy crude oil and water emulsion. Fuel. 2015;**141**:258-267

[37] Al-Sabagh AM, Nasser NM, Khamis EA. Demulsification of crude oil emulsions using ethoxylated aliphatic amine polyesters as novel demulsifiers. The International Journal of Science and Research. 2012;**3**(1171-83):17

[38] Dosher TM, Wise FA. Enhanced oil recovery potential. An estimate. Journal of Petroleum Technology. 1976;**SPE 5800**:575

[39] Green DW, Willhite GP, et al. Enhanced Oil Recovery. Vol. 6. Richardson, TX: Henry L. Doherty Memorial Fund of AIME, Society of Petroleum Engineers; 1998

[40] Kantzas A, Chatzis I, Dullien FAL, et al. Enhanced oil recovery by inert gas injection. In: SPE Enhanced Oil Recovery Symposium. Oklahoma: Society of Petroleum Engineers; 1988

[41] Nabipour M, Escrochi M, Ayatollahi S, Boukadi F, Wadhahi M, Maamari R, et al. Laboratory investigation of thermally-assisted gas–oil gravity drainage for secondary and tertiary oil recovery in fractured models. Journal of Petroleum Science and Engineering. 2007;**55**(1-2):74-82

[42] Santanna VC, Curbelo FDS, Castro Dantas TN, Dantas Neto AA, Albuquerque HS, Garnica AIC. Microemulsion flooding for enhanced oil recovery. Journal of Petroleum Science and Engineering. 2009;**66**(3-4):117-120

[43] Burauer S, Sachert T, Sottmann T, Strey R. On microemulsion phase behavior and the monomeric solubility of surfactant. Physical Chemistry Chemical Physics. 1999;**1**(18):4299-4306

[44] Perazzo A, Tomaiuolo G, Preziosi V, Guido S. Emulsions in porous media: From single droplet behavior to applications for oil recovery. Advances in Colloid and Interface Science. 2018. DOI: 10.1016/j.cis.2018.03.002

[45] Muggeridge A, Cockin A, Webb K, Frampton H, Collins I, Moulds T, et al. Recovery rates, enhanced oil recovery and technological limits. Philosophical Transactions of the Royal Society A. 2014;**372**(2006):20120320

[46] Strey HH, Parsegian VA, Podgornik R. Equation of state for polymer liquid crystals: Theory and experiment. Physical Review E. 1999;**59**(1):999

[47] Jakobs B, Sottmann T, Strey R, Allgaier J, Willner L, Richter D. Amphiphilic block copolymers as efficiency boosters for microemulsions. Langmuir. 1999;**15**(20):6707-6711

[48] Babadagli T. Analysis of oil recovery by spontaneous imbibition of surfactant solution. Oil & Gas Science and Technology. 2005;**60**(4):697-710

[49] Dantas TNC, Santanna VC, Neto AAD, Neto ELB, Moura MCPA.

Rheological properties of a new surfactant-based fracturing gel. Colloids and Surfaces A: Physicochemical and Engineering Aspects. 2003;**225**(1-3):129-135

[50] Castro Dantas TN, Santanna VC, Dantas Neto AA, Barros Neto EL. Application of surfactants for obtaining hydraulic fracturing gel. Petroleum Science and Technology. 2003;**21**(7-8):1145-1157

[51] Sharma T, Kumar GS, Sangwai JS. Enhanced oil recovery using oil-in-water (o/w) emulsion stabilized by nanoparticle, surfactant and polymer in the presence of NaCl. Geosystem Engineering. 2014;**17**(3):195-205

[52] Ki S, Friberg S. Microemulsions: Colloidal aspects. Advances in Colloid and Interface Science. 1975;**4**(4):281-300

[53] Prince MJA. Enhancing recovery by reducing surfactant adsorption on oil wet carbonate reservoirs, International Journal of ChemTech Research. 2017;**10**(7):715-722

[54] Austad T, Strand S. Chemical flooding of oil reservoirs. 4. Effects of temperature and pressure on the middle phase solubilization parameters close to optimum flood conditions. Colloids and Surfaces A: Physicochemical and Engineering Aspects. 1996;**108**(2-3):243-252

[55] Gurgel A, Moura MCPA, Dantas TNC, Neto ELB, Neto AAD. A review on chemical flooding methods applied in enhanced oil recovery. Brazilian Journal of Petroleum and Gas. 2008;**2**(2):18

[56] Yousufi MM, Elhaj MEM, Moniruzzaman M, Ayoub MA, Nazri ABM, Husin Hb, et al. Synthesis and evaluation of *Jatropha* oil-based emulsified acids for matrix acidizing of carbonate rocks. Journal of Petroleum Exploration and Production Technology. 2018:1-15. https://doi.org/10.1007/s13202-018-0530-8

[57] Siddiqui S, Nasr-El-Din HA, Khamees AA. Wormhole initiation and propagation of emulsified acid in carbonate cores using computerized tomography. Journal of Petroleum Science and Engineering. 2006;**54**(3-4):93-111